Classical and Quantum Information Theory for the Physicist

Classical and Quantum Information Theory for the Physicist

Harish Parthasarathy

Professor

Electronics & Communication Engineering

Netaji Subhas Institute of Technology (NSIT)

New Delhi, Delhi-110078

CRC Press
Taylor & Francis Group
Boca Raton London New York

CRC Press is an imprint of the
Taylor & Francis Group, an **informa** business

Manakin
PRESS

First published 2023
by CRC Press
4 Park Square, Milton Park, Abingdon, Oxon, OX14 4RN

and by CRC Press
6000 Broken Sound Parkway NW, Suite 300, Boca Raton, FL 33487-2742

CRC Press is an imprint of Informa UK Limited

The right of Harish Parthasarathy to be identified as author of this work has been asserted in accordance with sections 77 and 78 of the Copyright, Designs and Patents Act 1988.

Print edition not for sale in South Asia (India, Sri Lanka, Nepal, Bangladesh, Pakistan or Bhutan).

British Library Cataloguing-in-Publication Data
A catalogue record for this book is available from the British Library

ISBN: 9781032405179 (hbk)
ISBN: 9781032405209 (pbk)
ISBN: 9781003353454 (ebk)

DOI: 10.4324/9781003353454

Typeset in Arial, Minion Pro, Times New Roman, Wingdings, Calibri and Symbol
by Manakin Press, Delhi

Manakin
PRESS

Brief Contents

Detailed Contents

Preface

This book deals with certain important problems in classical and quantum information theory which is basically a subject involving the proof of coding theorems for data compression and the efficient transmission of information over noisy channels. By coding theorems, we mean proving the existence of codes that would achieve maximum possible data compression without distortion or with an admissible degree of distortion or would enable us to transmit data over a noisy channel so that the input message can be retrieved from the output to an arbitrary degree of accuracy (ie with arbitrarily small error probability) when the input data strings that are coded are made sufficiently long. In classical information theory, the important problems discussed in this book are as follows. First, the construction of no-prefix and uniquely decipherable codes from a probabilistic source and properties of the minimum average length of such codes with relation to the entropy of the source. Second, the proof the specific form of the Shannon entropy function based on fundamental properties that the entropy function of a source should possess like the conditional additivity of the joint entropy of two dependent sources, monotone increasing property of the entropy as a function of the number of alphabets in the source under a uniform distribution and continuity of the entropy as a function of the source probability distribution. For estabilishing properties of the minimum average code length, we require the Kraft inequality regarding the codeword lengths. The Kraft inequality is a relationship between the lengths of the codewords of the different source symbols and the size of the encoding alphabet. The validity of this inequality implies the existence of a no prefix code with the specified codeword lengths, the no prefix property of a code implies unique decipherability which in turn implies the validity of Kraft's inequality. The other fundamental result in the explicit construction of codes is optimal Huffman code which is a no prefix code of minimum average length. Then we come to the fundamental coding theorems of classical information theory namely Shannon's noiseless and noisy coding theorems. The former states that if data compression is carried out based on encoding long strings of iid or ergodic data, then the best achievable compression is when the code rate equals the source entropy in the case of iid data and the source entropy rate per symbol in the case of ergodic data. These theorems are established here using respectively the weak law of large numbers and the Shannon-Mcmillan-Breiman theorem the proof of which involves use of the ergodic theorem and the Martingale convergence theorem. These theorems show that when the source probability distribution is highly non-uniform, then significant data compression can be achieved, ie, if the source alphabet size is a, then without compression we require $a^n = 2^{n.log_2 a}$ memory places to store n-long data strings while with compression we require only $2^{n.H(X)}$ memory with negligible error probability during retrieval provided the data string length n is made sufficiently large. In other words, we are able to achieve data compression because $H(X) < log_2 a$ for a non-uniform source and hence $2^{nH(X)} << a^n$. Data compression is achieved because the weak law of large numbers enables us to neglect several data strings which have negligible proability of occurrence when the string length is made sufficiently large. This noiseless coding theorem of Shannon is thus a probabilistic result for it tells us that if we desire asymptotically zero distortion during decoding, then the maximal compression that one can achieve is obtained with an encoding rate of $H(X)/log_2 a < 1$, ie, when the n-length strings of the alphabet A of size a are encoded into $nH(X)/log_2 a$ length strings of the same alphabet or equivalently, when m-length binary strings that represent the iid source data are encoded into $mH(X)/log_2 a$ length binary strings. The compression per bit achieved equals $H(X)/log_2 a$ and the converse of this data compression theorem tells us further that we cannot do better. Shannon later on extended this result of his to reduce the data storage even further or equivalently increase the compression even further when a certain maximal distortion is allowed. This theory of Shannon is known as rate distortion theory. A simple proof of this result based on just the WLLN as in the case of the noiseless distortionless data compression theorem by exploiting properties of typicality and of sequences like for iid bivariate sequences $\{(X_i, W_i)\}$, the number of jointly typical sequences of length n is approximately $2^{nH(X,W)}$ if $\{W_i\}$ is given to be typical, then the number of $\{X_i\}$ sequences for which $\{(X_i, W_i)\}$ it typical is approximately $2^{H(X|W)}$, the probability of a typical sequence is approximately $2^{-nH(X)}$. Given a typical sequence $\{W_i\}$, the probability of $\{(X_i, W_i)\}$ being typical is approximately $2^{-nH(X|W)}$ the number of such conditionally typical sequences is approximately $2^{nH(X|W)}$. Given a typical $\{W_i\}$, assuming that $\{\tilde{X}_i\}$ has the same iid distribution as $\{X_i\}$ but is independent of $\{W_i\}$, the probability that $\{(\tilde{X}_i, W_i)\}$ is typical according to the distribution of $\{(X_i, W_i)\}$ is approximately $2^{-nH(X)}.2^{nH(X|W)} = 2^{-nI(X:W)}$.

The rate distortion theorem reduces compression per bit from $H(X)$ to $\min I(X : W) = H(X) - maxH(X|W)$ where the maximum is over all joint conditional distributions $p(W|X)$ subject to the average distortion condition $\sum_{X,W} d(X, W)p(X, W) \leq D$. In the case when $\{X_i\}$ is not necessarily iid but is ergodic, the same compression result holds but the simplest proof involves the use of large deviation theory. We've adapted our proof from the book by A.Dembo and O.Zeitouni on "Large deviations, Techniques and Applications". There are more recent results in noiseless data compression and rate distortion theory namely the extension of these results in the presence of available side information. Let $\{Y_i\}$ be the available iid side information so that X_i, Y_i are correlated for each i, but $\{(X_i, Y_i)\}$ is an iid bivariate sequence. Then, we can encode $\{X_i\}$ with a rate smaller than $H(X)$ and still obtain asymptotically zero distortion during the decoding process provided that the decoder makes use of the side information $\{Y_i\}$ for data recovery through its encoded form. In short, $\{X_i\}$ is encoded into an index i taking 2^{nR_1} values, $\{Y_i\}$ is encoded into an index s taking 2^{nR_2} values and the indices (i, s) are together used to decode $\{X_i\}$ correctly with asymptotic zero error probability provided that the code rate pair (R_1, R_2) falls within a region with the rate region for R_1 being much smaller than H(X). It should be noted that the rate region for (R_1, R_2) here is dictated by an auxiliary r.v U which tells us how much information about Y can be extracted for utilization in reducing the uncertainty in X. The resulting reduced uncertainty in X can then be completely removed by encoding $\{X_i\}$ at the appropriate reduced rate R_1.

The next important development involves using the side information in rate distortion theory to reduce the encoding rate even further than $minI(X : W)$, namely to $min(I(X : W) - I(Y : W))$ when $\{Y_i\}$ is the side information available so that X_i, Y_i are correlated but $\{(X_i, Y_i)\}$ is an iid bivariate sequence. The above minimum is over all conditional distributions $p(W|X)$ so that the joint distribution (Y, X, W) is $p(W|X)p(X, Y)$, ie, $Y \to X \to W$ forms a Markov chain and also over all decoders $\hat{X}^n = \{f(Y_i, W_i)\}$ with $p(W|X)$ and $f(Y, W)$ being subject to the maximum allowable distortion constraint $\mathbb{E}d(X, f(Y, W)) \leq D$. The idea in obtaining such a reduction in the encoding rate is that we use side information $Y^n = \{Y_i\}$ in the decoder to obtain successful decoding with allowable distortion D by partitioning the code index space $s \in \{1, 2, ..., 2^{nR_1}\}$ of the input sequence into 2^{nR_2} disjoint bins and showing that if the bins are selected appropriately, then the index of the bin is sufficient to yield the required estimate of the source word X^n upto a distortion level D. In proving this result, we make heavy use of the LLN in the form of properties of typical sequences and a fundamental result that if $Y^n \to X^n \to W^n$ is a Markov chain, then joint typicality of (X^n, W^n) implies joint typicality of (Y^n, X^n, W^n) and in particular, joint typicality of (Y^n, W^n). The number of bins 2^{nR_2} required for successful decoding is much smaller than 2^{nR_2} with $R_1 \approx I(X : W)$ which yields the desired reduction. Without the side information, Y^n available, we would not be able to decode with reduced rate because our method relies on decoding the s index in such a way that $(Y^n, W^n(s))$ is typical for some $s \in B(i)$ with the size of the bin $B(i)$ being approximately $2^{n(R_1-R_2)}$. As in the noiseless coding theorem and in all the other important results of information theory, it its the random coding argument that works here too, namely, our encoding for getting the index s and the bin index i are based on random codes for which we prove that the average error probability over all the random code ensembles converges to zero and hence there must exist one non random code for which the error probability converges to zero. The random coding argument is the classic approach initiated by Claude Shannon. We then prove again using Shannon's random coding argument the noisy coding theorem which states that for a DM channel specified by single symbol transition probabilities, if the rate of encoding is smaller than the channel capacity, then the decoding error probability converges to zero as the string size for encoding goes to infinity and conversely. The direct part of this theorem is true even for finite memory channels satisfying the m-independence condition but the converse is not true in general for non DM channels. All the converse results in information theory are proved using Fano's inequality which states that conditional uncertainity of the source message given the output is upper bounded by the number of source bits times the decoding error probability a and hence if the decoding error probability converges to zero, then the conditional uncertainty of the source message given the output per bit must also converge to zero. From this one obtains at once using properties of the joint and conditional entropy functions, the required upper bound on the maximum rate of information transmission for asymptotic error free decoding.

The other results that we prove here including the coverse are (a) distributed source coding in which given several correlated sources, if we encode each source sequence separately but jointly decode them exploiting their correlations, then we can compress to even less than the individual source entropies, namely to the conditional entropy of each source given the others. This is the famous Slepian-Wolf theorem. (b) Multiple access channel theory in which independent information form different sources is transmitted over multiple channels and the receiver a noise corrupted version of a function of these different inputs. Then the problem is to calculate the optimal rate region for the different message inputs for asymptotic error free decoding. (c) Broadcast channel in which multiple information from m sources arriving at rates $R_1, ..., R_m$ are encoded into a single input sequence which is received by several receivers m in number via respective channels connecting the source to their receivers after getting distorted by channel noise. The ith receiver must then decode the ith message from his received signal for each $i = 1, 2, ...m$. The objective is to determine the rate region here, ie the set of allowable rate m-tuples $(R_1, ..., R_m)$ for error free decoding at each of the m receivers. We determine the rate region here for the special case of a degraded broadcast channel in which each receiver receives noise corrupted signals only from their previous receivers in a sequence, ie, the set of all receivers is arranged in a sequence and each receiver receives signals only form the previous receiver while the first receiver receives signals from the channel connecting him directly to the source. Such a degraded broadcast channel puts severe restrictions on the nature of of the conditional probability distribution of the outputs given the input and this restriction makes it easy to evaluate the rate region. All these rate regions are evaluated as usual using the random coding argument. (d) Relay channels in which information goes from the main transmitter to the main receiver both directly and via an intermediate relay receiver and transmitter. The relay channel transmits additional bin information as earlier at a lower rate and hence the capacity region of the overall system can be enhanced as compared to the situation without relay. Finally, we discuss (e) the capacity region for a multi-terminal network with causal feedback which generalizes al the above mentioned situations, namelu that of multiple access, broadcast and relay channels. Here, there are m nodes with each node containing a transmitter, a receiver and an encoder. Each ordered pair of nodes is connected by a directed noisy channel, ie, any node can transmit signals to any other node. The encoded source symbol at time k at each node is a function of the messages to be transmitted from that node to all the other nodes as well as of the received signal at that node upto time $k - 1$. The decoder at node i of the message sent from the node j the given node i must be a function of the total received output at node i and the set of messages sent from node i to all the other nodes. The problem is to determine the size of the message set to be transmitted from each node to every other node so that error free decoding is achieved in the limit as the string size goes to infinity. The direct part of the achievability region does not have a proof while the converse part, namely necessary conditions on the rate region given that

The other topics covered in this book deal with the following kind of problems. (1) Design of quantum neural networks for tracking a given desired probability density function (pdf). Here we exploit the fact that since the generator of the Schrodinger evolution group is i times a Hermitian operator whenever the potential is a real valued function, it follows that the evolution operator is unitary at every time and hence the wave function at each time has a modulus square that integrates to unity so that the modulus square wave function is indeed a pdf. In other words, the Schrodinger equation naturally generates a family of pdf's parameterized by the time variable. This suggests that we can introduce a set of time varying control parameters into the potential with the time variation being controlled according to an adaptive algorithm so that the modulus square of the resulting wave function tracks the given pdf. This enables us to generate pdf's that are correlated with another signal/pdf by a training process. It finds application in predicting a given high dimensional pdf from a lower dimensional signal/pdf based on the trained Schrodinger potential. The idea here is to use the fact that the lower dimensional signal is correlated with the higher dimensional signal with the latter being modeled by the expected value of the position variable w.r.t. the modulus square of the wave function and to incorporate an error energy term in the adaptive dynamics of the control parameters involving the error square between the lower dimensional signal and a constant times the higher dimensional signal or more generally the prediction error energy square when the prediction is of the higher dimensional signal based on present and past samples of the lower dimensional signal. This error energy term is to be added to the error energy between the desired pdf and the modeled pdf obtained as the modulus square of the wave function. Such a quantum neural network also enables us to achieve data compression by compressing an entire one parameter family of pdf's into the time variation of a finite set of parameters. The connection between information theory and such a quantum neural network (qnn) is as follows. We can introduce random parameters into the Schrodinger equation and then assess how much information we have gained from the qnn by transforming a low dimensional signal/pdf into a high dimensional one by evaluating the Von-Neumann entropy of the average density operator at time t. This average density operator is $\rho(t) = \mathbb{E}(|\psi(t) >< \psi(t)|)$ where $|\psi(t) >$ is the wave function ket at time t. Equivalently in the position domain

$$\rho(t, q, q') = \mathbb{E}(\psi(t, q)\bar{\psi}(t, q')) = \int \psi(t, q, \omega)\psi(t, q', \omega)dP(\omega)$$

Even if random parameters are not introduced, we can calculate the empirical pdf of he parameters based on time averages:

$$p_\theta(\omega) = T^{-1} \int_0^T \delta(\omega - \theta(t))dt$$

and using this empirical pdf, evaluate the final density matrix as

$$\rho(t, , q') = \int \psi(t, q, \omega)\bar{\psi}(t, q', \omega)p_\theta(\omega)d\omega$$

where $\psi(t, q, \omega)$ is the wave function obained by solving the stationary Schrodinger equation with the parameters θ fixed at ω. In short, the time varying low dimensional pdf given as input to the qnn generates a rapid parameter variation which can be modeled as a random variable vector with a probability distribution that causes entropy to get pumped into the qnn. This gain in information while transforming a lower dimensional pdf to a higher dimensional one can therefore be assessed by computing the Von-Neumann entropy of the transformed state, by modeling the rapid parameter variations as a random parameter vector.

(2). The other kind of problem discussed in this book involves computing the entropy pumped by a quantum noisy bath into the dynamics of a quantum field or a quantum superstring. Here, the idea is first to set up the Hamiltonian of the field or the superstring in terms of Boson and Fermion creation and annihilation operators and then to consider the interaction Hamiltonian between this field or superstring and certain classical or quantum gauge fields. The dynamnics of these gauge fields can usually be expressed as superposition of the creation and annihilation processes in the Hudson-Parthasarathy quantum stochastic calculus theory. The dynamnics of the wave functional of the field or the superstring is then obtained using such a quantum noisy Hamiltonian in the form of the celebrated Hudson-Parthasarathy-noisy Schordinger equation and one can contract this equation by tracing out over the bath variables to obtain the GKSL master equation for the mixed state of the field or superstring state alone. From this mixed state, we compute the Von-Neumann entropy and then evaluate its rate of increase with time and hence derive conditions on the Lindblad operators that guarantee monotonic entropy increase in accordance with the second law of thermodynamics. (3).The next problem discussed here involves calculating the quantum corrections in powers of Planck's constant to the classical Fokker-Planck or more generally to the Kushner-Kallianpur stochastic filtering equations for the conditional probability density of the current state given measurements upto to the current time. To evaluate such corrections and make an analogy with the classical case, we must express all the quantum dynamical equations for the density matrix like the quantum Liouville equation the Belavkin filter equation etc. in terms of probability densities. However, in the quantum scenario, non commuting variables like position and momentum do not have joint probability densities as they are not simultaneously measurable and hence we must replace the quantum mixed state density matrix by the complex Wigner distribution function. The Wigner distribution function is complex function of the position and momentum variables whose marginals coincide with the probability densities of the position and momentum individually. Thus by transforming the quantum state/filtered state dynamics into the Wigner distribution language, we can get an idea of how much quantum corrections in powers of Planck's constant does quantum mechanics introduce into the evolution of a classical probability density. The classical joint probability density is herein replaced by its quantum analogue namely the Wigner distribution function which although is not a joint probability density, has properties similar to a joint probability density in that its marginals are probability densities and expected values of quantum observables in the mixed state can be evaluated in terms of integrals in position-momentum phase space using the Wigner distribution function. The crucial idea behind using the Wigner distribution in quantum mechanics to draw an analogy with the classical Fokker-Planck equation is that the Wigner distribution gives a complex function representation of the density operator which enables one to express the quantum laws of motion satisfied by the density operator as partial differential equations or stochastic partial differential equations satisfied by the Wigner distribution. Finally, we discuss a class of problems in quantum information theory that provide non-commutative generalizations of the corresponding problems in classical information theory and statistics. These problems are as follows. (1) The Cq channel coding theorem and its converse. Here, we have an input alphabet and strings of this input alphabet are encoded into corresponding tensor products of quantum states. The Cq channel is thus specified by the assignment of a density matrix to each input alphabet and hence a tensor product of density matrices/states to each input string. Such a Cq channel is thus seen to be a quantum generalization of the classical discrete memoryless channel. The memoryless property is characterized by the tensor product. Suppose that there are $N(n)$ input strings of length n to be transmitted. Then the asymptotic rate of transmission is $R = \lim \log(N(n))/n$. We introduce $N = N(n)$ detection operators $Y_1(n), ..., Y_N(n)$ which are all positive operators summing to the identity operator on the tensor product Hilbert space. If $\phi_k(n)$ is the input string transmitted corresponding to the message $k = 1, 2, ..., N(n)$, via the tensor product quantum state $W(\phi_k(n))$, then the probability at the received end of making the correct decision is $Tr(W(\phi_k(n))Y_k(n))$. The average decoding error probability is therefore $N(n)^{-1} \sum_{k=1}^{N(n)} Tr(W(\phi_k)(1 - Y_k(n)))$ and the problem is to determine conditions on the rate of transmission R so that this decoding error probability converges to zero as the string length n goes to infinity. The direct part of the Cq coding theorem states that a sequence of codes $(\phi_k(n))_{k=1}^{N(n)}, n = 1, 2, ...$ exists along with detection operators $(Y_k(n))_{k=1}^{N(n)}, n = 1, 2, ...$ for which the decoding error probability converges to zero iff $R < C$ where C is the Cq channel capacity defined as the maximum of the Cq mutual information of this channel, namely of $H(\sum_x p(x)W(x)) - \sum_x p(x)H(W(x))$ over all $p(.)$ where $W(x)$ is the quantum state into which the input alphabet x is encoded and $\{p(x)\}$ is a probability distribution over the input source alphabet. The direct part of this Cq coding theorem is proved using the random coding argument, ie, by choosing a random code $\phi(n) = \{\phi_k(n)\}_k$ whose components are iid with pdf $p(.)$, constructing detection operators as functions of this random code and proving that the average decoding error probability over all the random code ensembles converges to zero when $R < C$ and hence concluding the existence a non random code along with detection operators for which the error probability converges to zero. Here, we prove that the average decoding error probability is bounded by a quantity expressed in terms of the Cq relative Renyi entropy and by taking the limit of this relative Renyi entropy after noting that the limit of the relative Renyi entropy equals the quantum mutual Cq mutual information, we obtain the desired sufficiency result. For the converse, we upper bound the probability of no error occurring by a quantity similar to the relative Renyi entropy by making use of the monotonicity of the relative Renyi entropy under TPCP maps, ie, under quantum operations. We also supply a proof of this monotonicity of the relative Renyi entropy by making use of monotoncity of operator convex functions under TPCP maps and more generally under linear contraction maps on the space of matrices. The proof of the converse also makes use of an important matrix inequality, namely the reverse Holder inequality for positive matrices. Finally, we discuss the quantum binary hypothesis testing problem for tensor product states and prove the quantum Stein lemma in this context, namely that given that the false alarm probability converges to zero as the number of tensor components goes to infinity, the probability of missing the target will converge to zero at an exponential rate equal to the negative of the relative entropy between the two states and we cannot do better. The quantum Stein lemma is thus a non-commutative generalization of the classical Stein lemma involving the asymptotic rates of the error probabilities for an infinite sequence of iid random variables. In this book, we also discuss some examples of Cq channels arising in quantum filtering theory like a source alphabet being encoded into the initial system state which gets transmitted over a quantum noisy channel specified by the Hudson-Parthasarathy noisy Schrodinger equation along with non-demolition measurements and from these measurements at the received end, we have by application of the Belavkin filter an evolving estimate of the quantum state on which by taking repeated measurements, we wish to extract the input alphabet that was encoded in to the initial quantum state. Repeated measurements cause the state collapse postulate to be applicable so that in between measurements, the filtered state evolves according to the Belavkin quantum filter and just after each measurement, the state gets collapsed.

Author

Chapter 1

Quantum Information Theory, A Selection of Matrix Inequalities

1.1 Monotonicity of quantum relative Renyi Entropy

Statement of the result: Let $0 \leq s \leq 1$ and let K be a TPCP map acting on the convex set of states in a finite dimensional Hilbert space. Then, for any two states $+\rho, \sigma$, we have

$$Tr(K(\sigma)^s K(\rho)^{1-s}) \geq Tr(\sigma^s \rho^{1-s})$$

More generally, let f be an operator convex function. Define the linear operator $\Delta_{\rho,\sigma}$ on the vector space of matrices by

$$\Delta_{\rho,\sigma} = R_\rho L_\sigma^{-1}$$

where

$$R_Y(X) = XY, L_Y(X) = YX$$

Thus,

$$\Delta_{\rho,\sigma}(X) = \sigma^{-1} X \rho$$

Note that for any two matrices X, Y, R_X and L_Y commute:

$$R_X L_Y(Z) = L_Y R_X(Z) = YZX$$

Then, we have the inequality
Define the linear operator $K_{\sigma,r}$ on the space of matrices by

$$K_{\sigma,r}(X) = K(\sigma)^{-1}(K(\sigma X)$$

Define the following inner products on the space of complex matrices of appropriate size:

$$< Y, X >_{K(\sigma)} = Tr(Y^*K(\sigma)X)), < Y, X >_\sigma = Tr(Y^*\sigma X)$$

Note that the sizes of the matrices in both the cases will generally differ. For example if K is the partial trace operation Tr_2 on $\mathbb{C}^{n \times n} \otimes \mathbb{C}^{m \times m}$ then in the first case, Y, X will be in $\mathbb{C}^{n \times n}$, while in the second case, they will be in $\mathbb{C}^{nm \times nm}$. Now, let K be the partial trace operation as defined here. Then for $Y \in \mathbb{C}^{n \times n}$ and $X \in \mathbb{C}^{nm \times nm}$,

$$< Y, K_{\sigma,r}(X) >_{K(\sigma)} = Tr(Y^*K(\sigma X)) = Tr(Y^*.Tr_2(\sigma X))$$

$$= Tr((Y^* \otimes I)\sigma X) = < Y \otimes I, X >_\sigma$$

This means that relative to the inner products $< .,. >_{K(\sigma)}$ and $< .,. >_\sigma$ on $\mathbb{C}^{n \times n}$ and $\mathbb{C}^{nm \times nm}$ respectively, the adjoint of $K_{\sigma,r}$ is given by $K^*_{\sigma,r}$ where

$$K^*_{\sigma,r}(Y) = Y \otimes I$$

ie

$$< Y, K_{\sigma,r}(X) >_{K(\sigma)} = < K^*_{\sigma,r}(Y), X >_\sigma$$

Further,

$$K_{\sigma,r}\Delta_{\rho,\sigma}K^*_{\sigma,r}(Y) =$$

$$= K_{\sigma,r}\Delta_{\rho,\sigma}(Y \otimes I) =$$

$$K(\sigma)^{-1}K(\sigma\Delta_{\rho,\sigma}(Y \otimes I))$$

$$= K(\sigma)^{-1}K((Y \otimes I)\rho) = K(\sigma)^{-1}Tr_2((Y \otimes I)\rho)$$

$$= K(\sigma)^{-1}Y.Tr_2(\rho) = K(\sigma)^{-1}YK(\rho) = \Delta_{K(\rho),K(\sigma)}(Y)$$

Thus, when K is *the* Tr_2 operation (partial trace), then

$$K_{\sigma,r}\Delta_{\rho,\sigma}K^*_{\sigma,r} = \Delta_{K(\rho),K(\sigma)}$$

Therefore from the operator convexity of f,

$$f(\Delta_{K(\rho),K(\sigma)}) \leq K_{\sigma,r}f(\Delta_{\rho,\sigma})K^*_{\sigma,r}$$

where K is the partial trace operation. This means that for any matrix X,

$$< X, f(\Delta_{K(\rho),K(\sigma)})(X) >_{K(\sigma)} \leq$$

$$< X, K_{\sigma,r}f(\Delta_{\rho,\sigma})K^*_{\sigma,r}(X) >_{K(\sigma)}$$

$$= < K^*_{\sigma,r}(X), f(\Delta_{\rho,\sigma})K^*_{\sigma,r}(X) >_\sigma$$

$$= < X \otimes I, f(\Delta_{\rho,\sigma})(X \otimes I) >_\sigma$$

or equivalently,

$$Tr(X^*K(\sigma)f(\Delta_{K(\rho),K(\sigma)})(X)) \leq$$

$$Tr((X^* \otimes I)\sigma . f(\Delta_{\rho,\sigma})(X \otimes I))$$

In general, if K is any TPCP map, it has the Stinespring representation

$$K(\rho) = Tr_2(U(\rho \otimes \rho_0)U^*)$$

where ρ_0 is a state in a different Hilbert space and U is a unitary operator. In this case, application of the above result gives

$$Tr(X^* K(\sigma) f(\Delta_{K(\rho),K(\sigma)})(X))$$

$$= Tr(X^* Tr_2(U(\sigma \otimes \rho_0)U^*) f(\Delta_{Tr_2(U(\rho \otimes \rho_0)U^*),Tr_2(U(\sigma \otimes \rho_0)U^*)})(X))$$

$$\leq Tr((X^* \otimes I)(U(\sigma \otimes \rho_0)U^*) f(\Delta_{U(\rho \otimes \rho_0)U^*,U(\sigma \otimes \rho_0)U^*})(X \otimes I))$$

Taking

$$f(x) = -x^{1-s}, 0 \leq s \leq 1$$

and observing that

$$< Y, f(\Delta_{A,B})(X) >_B = Tr(Y^* B f(\Delta_{A,B}(X)) =$$

$$-Tr(Y^* B^s X A^{1-s})$$

gives us the monotonicity of the quantum relative entropy under any TPCP map K:

$$Tr(X^* K(\sigma)^s X K(\rho)^{1-s}) \geq Tr((X^* \otimes I)U(\sigma^s \otimes \rho^s)U^*(X \otimes I)U(\rho^{1-s} \otimes \rho_0^s)U^*)$$

for any matrix X of appropriate size and therefore, taking $X = I$ gives us

$$Tr(K(\sigma)^s K(\rho)^{1-s}) \geq Tr(\sigma^s \rho^{1-s}), 0 \leq s \leq 1$$

1.2 Problems

[1] Show that if Q is a doubly stochastic matrix, then it can be expressed as a convex linear combination of permutation matrices $U_k, k = 1, 2, ..., r$, ie

$$Q = \sum_{k=1}^{r} t(k)U_k, t(k) > 0, \sum_k t(k) = 1$$

Thus, if p is any probability distribution on the states and H the entropy map, then

$$H(Q(p)) = H(\sum_k t(k)U_k p) \geq \sum_k t(k)H(U_k p) = H(p)$$

since if U is any permutation, then $H(Up) = H(p)$.

[2] This problem deals with the relationship between a TPCP map acting on a mixed quantum state and the corresponding stochastic matrix acting on the probability vector that forms the set of eigenvalues of the quantum state.

Let K be a TPCP map in a Hilbert space. Then, if ρ is a state with spectral decomposition

$$\rho = \sum_k |e_k > p(k) < e_k|$$

then

$$\sigma = K(\rho) = \sum_k p_k K(|e_k >< e_k|)$$

Writing the spectral decomposition of σ as

$$\sigma = \sum_k |f_k > q(k) < f_k|$$

it follows that

$$q(k) = \sum_j < f_k|K(|e_j >< e_j|)|f_k > p_j = \sum_j Q_{kj} p_j$$

where

$$Q_{kj} =< f_k|K(|e_j >< e_j|)|f_k >, k, j = 1, 2, ..., n$$

It is easily seen that

$$\sum_k Q_{kj} = Tr(K(|e_j >< e_j|)) = Tr(|e_j >< e_j|) = 1$$

$$\sum_j Q_{kj} =< f_k|K(I)|f_k >$$

This need not be one in general but it is one if K is a pinching, ie, a PVM and not only a POVM. In that case Q is a doubly stochastic matrix and it follows that

$$H(\sigma) = H(K(\rho)) = H(q) = H(Q(p)) \geq H(p) = H(\rho)$$

which proves the monotonicity of quantum entropy under pinching operations.

Chapter 2

Stochastic Filtering Theory Applied to Electromagnetic Fields and Strings

2.1   M.Tech dissertation topics

1.Filtering theory for quantum strings.
   2.Verification of quantum coding theorems using MATLAB.
   3.Filtering theory for fluid velocity fields in curved space-time.
   4.Quantum neural networks using mixed state evolution.

2.2   Estimating the time varying permittivity and permeability of a region of space using non-linear stochastic filtering theory

The Maxwell equations are

$$curl E = -(\mu(t,r)H(t,r))_{,t}, curl H = J(t,r) + (\epsilon(t,r)E(t,r))_{,t}$$
$$div(\epsilon(t,r)E) = 0, div(\mu(t,r)H) = 0$$

From these equations, we derive

$$-\nabla^2 E + \nabla(div E) = -\mu_{,t}(J + (\epsilon E)_{,t}) - \nabla(\mu_{,t}) \times H - (\nabla\mu) \times H_{,t} - \mu(J_{,t} + (\epsilon E)_{,tt})$$

$$-\nabla^2 H + \nabla(div H) = curl J + \epsilon_{,t}(\mu H)_{,t} + \nabla(\epsilon_{,t}) \times E + (\nabla\epsilon) \times E_{,t} - \epsilon(\mu H)_{,tt}$$

$$\epsilon . div E + (\nabla\epsilon, E) = 0, \mu . div H + (\nabla\mu, H) = 0$$

or equivalently,

$$div E = -(\nabla log(\epsilon), E), div H = -(\nabla log(\mu), H)$$

5

writing

$$\epsilon(t,r) = \epsilon_0(1 + \sum_{k=1}^{p} \theta_k(t)f_k(r), \mu(t,r) = \mu_0(1 + \sum_{k=1}^{p} \theta_k(t)g_k(r))$$

where $\theta_k(t)'s$ are of the first order of smallness, we find that these equations can be expressed as

$$(\nabla^2 - \partial_t^2)\psi(t,r) = \sum_k (\theta_k(t)(A_k(r)\psi + B_k(r)(\nabla \otimes \psi)) + F_0(t,r) + \sum_k \theta_k(t)F_{1k}(t,r) + \theta_k'(t)$$

$$+ \sum_k (\theta_k''(t)C_k(r)\psi(t) + \theta_k'(t)D_k(r)\psi'(t) + \theta_k(t)E_k(r)\psi''(t))$$

where A_k, B_k, C_k, D_k, E_k are matrices of appropriate size that depend only on the spatial coordinates. F_0, F_{1k}, F_{2k} are source functions depending on the current density J and its first order space-time derivatives. Here,

$$\psi(t,r) = [E(t,r)^T, H(t,r)^T]^T$$

is a six dimensional vector valued function of the space-time coordinates. Here, second order of smallness terms in the functions θ_k and its derivatives have been neglected. Again, after some manipulations it can be brought to the standard form

$$\partial_t^2 \psi(t,r) = \nabla^2 \psi(t,r) + \sum_k \theta_k(t)E_k(r)\nabla^2\psi(t,r)$$

$$+ \sum_k (\theta_k(t)(A_k(r)\psi + B_k(r)(\nabla \otimes \psi)) + F_0(t,r)$$

$$+ \sum_k \theta_k(t)F_{1k}(t,r) + \theta_k'(t)F_{2k}(t,r)) + \sum_k (\theta_k''(t)C_k(r)\psi(t) + \theta_k'(t)D_k(r)\psi'(t)$$

and we are now in a position to apply stochastic filtering theory to this equation after defining a measurement model for estimating the functions $\theta_k(t)$ on a real time basis. The stochastic dynamics of the $\theta_k(t)'s$ must however first be specified.

2.3　Filtering theory for quantum superstrings

The Bosonic string has the Lagrangian

$$L_B = (1/2)\partial_\alpha X^\mu \partial^\alpha X_\mu$$

and the Fermionic string has the Lagrangian

$$L_F = -i\bar{\psi}\rho^\alpha \partial_\alpha \psi$$

where
$$\bar{\psi} = \psi^T \rho^0$$

with
$$\rho^0 = \sigma_2, \rho^1 = i\sigma_1$$

Then
$$(\rho^0)^2 = I, \rho^0 \rho^1 = \sigma_3$$

and hence, writing
$$\psi = [\psi_-, \psi_+]^T$$

we get using
$$\rho^0 (\rho^0 \partial_0 + \rho^1 \partial_1)$$
$$= \partial_0.I + \sigma_3 \partial_1$$

that
$$L_F = -i(\psi_- \partial_+ \psi_- + \psi_+ \partial_- \psi_+)$$

where
$$\partial_+ = \partial_0 + \partial_1,$$
$$\partial_- = \partial_0 - \partial_1$$

Thus, the Fermionic string equations are
$$\partial_+ \psi_- = 0, \partial_- \psi_+ = 0$$

These have solutions
$$\psi_- = \sum_n S_n^- .exp(in(\tau - \sigma))$$
$$\psi_+ = \sum_n S_n^+ exp(in(\tau + \sigma_3))$$

The Fermionic Lagrangian is given by
$$L_F = -i \int_{-\pi}^{\pi} (\psi_- \partial_+ \psi_- + \psi_+ \partial_- \psi_+) d\sigma$$

and this vanishes when the equations of motion hold. The canonical momenta are
$$P_- = \delta L_F / \delta \partial_0 \psi_- = \psi_-,$$
$$P_+ = \delta L_F / \delta \partial_0 \psi_+ = \psi_+$$

So the Hamiltonian density is
$$H_F = P_- \partial_0 \psi_- + P_+ \partial_0 \psi_+ - L_F$$
$$= -\psi_- \partial_1 \psi_- + \psi_+ \partial_1 \psi_+$$

and when the equations of motion hold, this equals
$$H_F = \psi_- \partial_- \psi_- + \psi_+ \partial_+ \psi_+$$

and the Hamiltonian is given by

$$H_F = \int_{-\pi}^{\pi} (\psi_- \partial_- \psi_- + \psi_+ \partial_+ \psi_+) d\sigma$$

$$= \sum_n n(S_n^-.S_{-n}^- + S_n^+.S_{-n}^+)$$

The canonical anticommutation relations are

$$[\psi_-^a(\tau,\sigma), P_-^b(\tau,\sigma')]_+ = i\delta(\sigma - \sigma')$$

$$[\psi_+^a(\tau,\sigma), P_+^b(\tau,\sigma')]_+ = i\delta_{ab}\delta(\sigma - \sigma')$$

and these result in

$$[\psi_-^a(\tau,\sigma), \psi_-^b(\tau,\sigma')]_+ =$$
$$\delta_{ab}\delta(\sigma - \sigma'),$$
$$[\psi_+^a(\tau,\sigma), \psi_+^b(\tau,\sigma')]_+ =$$
$$\delta_{ab}\delta(\sigma - \sigma')$$

Thus,

$$[S_n^{-a}, S_m^{-b}]_+ = \delta_{ab}\delta(n + m),$$
$$[S_n^{+a}, S_m^{+b}]_+ = \delta_{ab}\delta(n + m),$$

and of course

$$[S_n^{-a}, S_m^{+b}]_+ = 0$$

Note that actually ψ is the shorthand notation for $\psi^{\mu a}$. in other words, ψ has a vector index as well as a spinor index. Thus the Fermionic Lagrangian is actually

$$\bar{\psi}^\mu \rho^\alpha \partial_\alpha \psi_\mu$$

$$= \psi_-^{\mu T} \partial_+ \psi_{-\mu}$$

$$+ \psi_+^{\mu T} \partial_- \psi_{+\mu}$$

The supersymmetry transformations that leave the sum of the Bosonic and Fermionic actions invariant are

$$\delta X^\mu = \bar{\epsilon}\psi^\mu,$$

$$\delta \psi^\mu = \rho^\alpha \epsilon \partial_\alpha X^\mu$$

Where ϵ is an infinitesimal Fermionic parameter. We now add supesymmetry breaking terms by introducing current and vector potential sources just as the charge and current sources add interaction terms to the electromagnetic field action and the vector potential sources add interaction terms to the Dirac action for the electron-positron field. These source terms are

$$eA_\alpha \bar{\psi}\rho^\alpha \psi,$$

and

$$J^\alpha_\mu \partial_\alpha X^\mu$$

After addition of these, the superstring equations become

$$\Box X^\mu = J^\alpha_{\mu,\alpha},$$

$$i\rho^\alpha \partial_\alpha \psi = eA_\alpha \rho^\alpha \psi$$

Another kind of supersymmetry breaking term arises when the Fermionic current field $\bar\psi \rho^\alpha \psi$ interacts with the bosonic string field. The resulting interaction term is given by

$$f_\mu \bar\psi \rho^\alpha \psi . \partial_\alpha X_\mu$$

where $f_\mu(\tau, \sigma)$ is an external random source field.

Superymmetry current: Consider a local supersymmetry transformation:

$$\delta X^\mu = \bar\epsilon \psi^\mu,$$

$$\delta \psi^\mu = \rho^\alpha \epsilon \partial_\alpha X^\mu$$

where now the supersymmetry Fermionic parameter ϵ is local, ie, it depends on (τ, σ). Under such a transformation, the Bosonic Lagrangian changes by

$$\partial^\alpha X^\mu \partial_\alpha \delta X_\mu$$

and this contains a term involving the derivatives of ϵ. This term is given by

$$\partial^\alpha X^\mu (\partial_\alpha \bar\epsilon) \psi_\mu$$

Likewise the change in the Fermionic Lagrangian contains a term

$$\bar\psi^\mu \rho^\alpha \rho^\beta (\partial_\beta \epsilon) \partial_\alpha X_\mu$$

The other terms in the sum of the Bosonic and Fermionic Lagrangian that do not involve derivatives of ϵ cancel out after integration over the $\tau - \sigma$ space which is precisely the manifestation of global supersymmetry. The equations of motion imply therefore that the sum of the terms involving derivatives of ϵ after integration should cancel out and this is precisely the condition of conservation of the supersymmetry current when the equations of motion hold:

$$\partial_\beta J^\beta = 0$$

where

$$J^\beta = (\partial_\alpha X_\mu) \rho^\alpha \rho^\beta \psi^\mu$$

or equivalently,

$$J^\alpha = (\partial_\beta X_\mu) \rho^\beta \rho^\alpha \psi^\mu$$

We can include a term in the superstring Lagrangian involving the interaction of this supersymmetry current with a stochastic c-number field and then write down the equations of motion. Note that for each $\alpha = 0, 1$, J^α is a two component spinor and hence such a coupling term will have the form

$$\Delta L = \bar{B}_\alpha J^\alpha = B_\alpha^T \rho^0 J_\alpha$$

The resulting modified equations of motion are

$$\Box X^\mu = \partial_\beta (\bar{B}_\alpha \rho^\beta \rho^\alpha \psi^\mu),$$

$$\rho^\alpha \partial_\alpha \psi^\mu = \rho^\alpha \rho^\beta \bar{B}_\alpha \partial_\beta X^\mu$$

Note that here $B_\alpha(\tau, \sigma)$ is a stochastic spinor field. We are now in a position to evaluate the magnitude of the supersymmetry breaking terms in the equations of motion. The supersymmetry breaking term in the Bosonic equation of motion is given by

$$\partial_\beta (\bar{B}_\alpha \rho^\beta \rho^\alpha \psi^\mu)$$

In this equation, we substitute in ψ^μ its expression corresponding to the free Fermionic field

$$\psi^{a\mu}(\tau, \sigma) = \sum_n S^{a\mu}(n).exp(in(\tau - \sigma))$$

where

$$[S^{a\mu}(n), S^{b\nu}(m)]_+ = \delta_{ab} \eta_{\mu\nu} \delta(n + m)$$

and then compute its mean square value in a Fermionic Coherent state. We have

$$\bar{B}_\alpha \rho^\beta \rho^\alpha \psi^\mu =$$

$$\bar{B}_\alpha \rho^\beta \rho^\alpha \sum_n S^\mu(n) e_n(\tau - \sigma)$$

where

$$e_n(\tau) = exp(in\tau)$$

Then,

$$f^\mu = \partial_\beta (\bar{B}_\alpha \rho^\beta \rho^\alpha \psi^\mu)$$

$$= \bar{B}_{\alpha,\beta} \rho^\beta \rho^\alpha \sum_n S^\mu(n) e_n(\tau - \sigma)$$

$$+ i \bar{B}_\alpha \rho^\beta \rho^\alpha u_\beta \sum_n n S^\mu(n) e_n(\tau - \sigma)$$

where

$$((u_\alpha)) = (1, -1)^T$$

The average value of $f^\mu \partial_0 \bar{X}_\mu$ over the string gives us the "four power" pumped into the string by the supersymmetry breaking forces. Writing

$$X_\mu(\tau, \sigma) = -i \sum_n (a_\mu(n)/n) e_n(\tau - \sigma)$$

gives

$$\partial_0 X_\mu = \sum_n a_\mu(n) e_n(\tau - \sigma)$$

and therefore we evaluate

$$(2\pi)^{-1} \int_{-\pi}^{\pi} f^\mu(\tau, \sigma) \partial_0 X_\mu(\tau, \sigma) d\sigma$$

$$= \bar{B}_{\alpha,0}(\tau) \rho^0 \rho^\alpha \sum_n S^\mu(n) a_\mu(-n)$$

$$+ i \bar{B}_\alpha(\tau) \rho^\beta \rho^\alpha u_\beta \sum_n n S^\mu(n) a_\mu(-n)$$

assuming that B_α is a function of only τ, an approximation which states that the stochastic field B_α does not vary too rapidly over the length scale of the string. To evaluate the quantum average of this supersymmetry breaking four power, we have to compute the quantum average

$$< S^{a\mu}(n) a_\mu(-n) >$$

in a coherent state of the Bosons and Fermions. To evaluate the quantum average of the square of the four power, we have to evaluate the quantum average

$$< S^{a\mu}(n) S^{b\nu}(m) a_\mu(-n) a_\nu(-m) >$$

in a coherent state of the Bosons and Fermions. Assume that $|\phi(u) >$ is a coherent state of the Bosons so that

$$a_\mu(n)|\phi(u) >= u_\mu(n)|\phi(u) >, n > 0,$$

and that $|\psi(v) >$ is a coherent state of the Fermions so that

$$S^{a\mu}(n)|\psi(v) >= v^{a\mu}(n)|\psi(v) >$$

Problem: Write down the classical superstring string equations in the presence of an external string gauge field with random forces incorporated into the gauge field and by taking discrete measurements at a finite set of points on the string over a continuous time interval, develop the Kushner-Kallianpur stochastic filtering equations for estimating the string field at all points. Generalize this to the quantum superstring case by setting up the Hamiltonian of the superstring in terms of the Bosonic and Fermnionic string creation and annihilation operators and assuming the quantum noisy gauge field to be given by a superposition of creation and annihilation processes in the sense of Hudson and Parthasarathy with coefficients dependent on the superstring or equivalently on the Bosonic and Fermionic creation and annihilation operators.

2.4 Study project:Reduction of supersymmetry breaking by feedback

Problem: Write down the superstring equations in the presence of supersymmetry breaking Lagrangian terms. Then write down the superstring equations in the absence of supersymmetry breaking terms. Apply feedback forces to the former set of equations with the feedback terms being based on the error between the superstring values of the two solutions at a discrete set of spatial string points with the feedback forces being designed to reduce the discrepancy between the two solutions. Now in the broken supersymmetry dynamical equations, take random forces into account and obtain EKF based filtered estimates of the superstring field values based on noisy output data collected at a discrete set of string points and for feedback, use the error between the desired unbroken supersymmetry field solutions and the EKF based estimate of the broken supersymmetry field with the feedback coefficients being designed to reduce the error.

Chapter 3

Wigner-distributions in Quantum Mechanics

3.1 Quantum Fokker-Planck equation in the Wigner domain

Since Q, P do not commute in quantum mechanics, we cannot talk of a joint probability density of these two at any time unlike the classical case. So how to interpret the Lindblad equation

$$i\partial_t \rho = [H, \rho] - (1/2)(L^* L \rho + \rho L^* L - 2L\rho L^*)$$

as a quantum Fokker-Planck equation for the joint density of (Q, P) ? Wigner suggested the following method: Let $\rho(t, Q, Q') = <Q|\rho(t)|Q>$ denote $\rho(t)$ in the position representation. Then $\rho(t)$ in the momentum representation is given by

$$\hat{\rho}(t, P, P') = <P|\rho(t)|P'> = \int <P|Q><Q|\rho(t)|Q'><Q'|P'> dQdQ'$$

$$= \int \rho(t, Q, Q') exp(-iQP + iQ'P') dQdQ'$$

$\rho(t, Q, Q)$ is the probability density of $Q(t)$ and $\hat{\rho}(t, P, P)$ is the probability density of $P(t)$. Both are non-negative and integrate to unity. Now Wigner defined the function

$$W(t, Q, P) = \int \rho(t, Q + q/2, Q - q/2) exp(-iPq) dq/(2\pi)$$

This is in general a complex function of Q, P and therefore cannot be a probability density. However,

$$\int W(t, Q, P) dP = \rho(t, Q, Q),$$

is the marginal density of $Q(t)$ Further, we can write

$$W(t, Q, P) = \int \hat{\rho}(t, P_1, P_2) exp(iP_1(Q+q/2) - iP_2(Q-q/2)) exp(-iPq) dP_1 dP_2 dq / (2$$

$$- = \int \hat{\rho}(t, P_1, P_2) exp(i(P_1 - P_2)Q) \delta((P_1 + P_2)/2 - P) dP_1 dP_2 / (2\pi)$$

so that

$$\int W(t, Q, P) dQ = \int \hat{\rho}(t, P_1, P_2) \delta(P_1 - P_2) \delta((P_1 + P_2)/2 - P) dP_1 dP_2$$

$$= \hat{\rho}(t, P, P)$$

is the marginal density of $P(t)$. This suggest strongly that the complex valued function $W(t, Q, P)$ should be regarded as the quantum analogue of the joint probability density of $(Q(t), P(t))$. Our aim is to derive the pde satisfied by W when $\rho(t)$ satisfies the quantum Fokker-Planck equation. To this end, observe that

$$[P^2, \rho]\psi(Q) = (-\int \rho_{,11}(Q, Q')\psi(Q')dQ' + \int \rho(Q, Q')\psi''(Q')dQ'$$

$$= \int (-\rho_{,11} + \rho_{,22})(Q, Q')\psi(Q')dQ'$$

which means that the position space kernel of $[P^2, \rho]$ is

$$[P^2, \rho](Q, Q') = (-\rho_{,11} + \rho_{,22})(Q, Q')$$

Also

$$[U, \rho](Q, Q') = (U(Q) - U(Q'))\rho(Q, Q')$$

Let the Lindblad operator be given by

$$L = aQ + bP, a, b \in \mathbb{C}$$

Then,

$$L^* L\rho(Q, Q') = (\bar{a}Q + \bar{b}P)(aQ + bP)\rho(Q, Q')$$
$$= |a|^2 Q^2 \rho(Q, Q') - |b|^2 \rho_{,11}(Q, Q') - i\bar{a}bQ\rho_{,1}(Q, Q')$$
$$- ia\bar{b}(\rho(Q, Q') + Q\rho_{,1}(Q, Q'))$$

How to transform from $W(Q, P)$ to $\rho(Q, Q')$?

$$W(Q, P) = \int \rho(Q + q/2, Q - q/2) exp(-iPq) dq$$

gives

$$\int W(Q, P) exp(iPq) dP / (2\pi) = \rho(Q + q/2, Q - q/2)$$

or equivalently,

$$\int W((Q+Q')/2, P)exp(iP(Q-Q'))dP/(2\pi) = \rho(Q,Q')$$

So

$$\rho_{,1}(Q,Q') = \int ((1/2)W_{,1}((Q+Q')/2, P)+$$
$$iPW((Q+Q')/2, P))exp(iP(Q-Q'))dP$$

Also note that

$$Q\rho(Q,Q') = \int QW((Q+Q')/2, P)exp(iP(Q-Q'))dP =$$

$$= \int ((Q+Q')/2) + (Q-Q')/2)W((Q+Q')/2, P)exp(iP(Q-Q'))dP$$

$$\int (((Q+Q')/2)W((Q+Q')/2, P) + (i/2)W_{,2}((Q+Q')/2, P))exp(iP(Q-Q'))dP$$

on using integration by parts. Consider now the term

$$(U(Q) - U(Q'))\rho(Q,Q')$$

In the classical case, $\rho(Q,Q')$ is diagonal and so such a term contributes only when Q' is in the vicinity of Q. For example, if we have

$$\rho(Q,Q') = \rho_1(Q)\delta(Q' - Q) + \rho_2(Q)\delta'(Q' - Q)$$

then,

$$(U(Q') - U(Q))\rho(Q,Q') = \rho_2(Q)(U(Q') - U(Q))\delta'(Q' - Q)$$

or equivalently,

$$\int (U(Q') - U(Q))\rho(Q,Q')f(Q')dQ' =$$
$$-\partial_{Q'}(\rho_2(Q)(U(Q') - U(Q))f(Q'))|_{Q'=Q}$$
$$= -\rho_2(Q)U'(Q)f(Q)$$

which is equivalent to saying that

$$(U(Q') - U(Q))\rho(Q,Q') = -\rho_2(Q)U'(Q)\delta(Q' - Q)$$

just as in the classical case. In the general quantum case, we can write

$$\int \rho(Q,Q')f(Q')dQ' = \sum_{n\geq 0} \rho_n(Q)\partial_Q^n f(Q)$$

This can be seen by writing

$$f(Q') = \sum_n f^{(n)}(Q)(Q' - Q)^n/n!$$

and then defining

$$\int \rho(Q,Q')(Q'-Q)^n dQ'/n! = \rho_n(Q)$$

Now, this is equivalent to saying that the kernel of ρ in the position representation is given by

$$\rho(Q,Q') = \sum_n \rho_n(Q)\delta^{(n)}(Q-Q')$$

and hence

$$\int (U(Q')-U(Q))\rho(Q,Q')f(Q')dQ'$$

$$= \sum_n \rho_n(Q) \int \delta^{(n)}(Q-Q')(U(Q')-U(Q))f(Q')dQ'$$

$$= \sum_n \rho_n(Q)(-1)^n \partial_{Q'}^n((U(Q')-U(Q))f(Q'))|_{Q'=Q}$$

$$= \sum_{n\geq 0, 1\leq k\leq n} \rho_n(Q)(-1)^n \binom{n}{k}U^{(k)}(Q)f^{(n-k)}(Q)$$

Thus,

$$(U(Q')-U(Q))\rho(Q,Q') =$$

$$\sum_{n\geq 0, 1\leq k\leq n} \rho_n(Q)(-1)^k \binom{n}{k}U^{(k)}(Q)\delta^{(n-k)}(Q'-Q)$$

This is the quantum generalization of the classical term

$$-U(Q)\partial_P f(Q,P)$$

which appears in the classical Fokker-Planck equation. Another way to express this kernel is to directly write

$$(U(Q')-U(Q))\rho(Q,Q') = \sum_{n\geq 1} U^{(n)}(Q)(Q'-Q)^n \rho(Q,Q')/n!$$

From the Wigner-distribution viewpoint, it is more convenient to write

$$U(Q')-U(Q) = U((Q+Q')/2-(Q-Q')/2) - U((Q+Q')/2+(Q-Q')/2)$$

$$= -\sum_{k\geq 0} U^{(2k+1)}((Q+Q')/2)(Q-Q')^{2k+1)}/(2k+1)!2^{2k}$$

Consider a particular term $U^{(n)}((Q+Q')/2)(Q-Q')^n \rho(Q,Q')$ in this expansion:

$$U^{(n)}((Q+Q')/2)(Q-Q')^n \rho(Q,Q')$$

$$= U^{(n)}((Q+Q')/2)(Q-Q')^n \int W((Q+Q')/2,P)exp(iP(Q-Q'))dP$$

$$= i^n \int U^{(n)}((Q+Q')/2)\partial_P^n(W((Q+Q')/2,P))exp(iP(Q-Q'))dP$$

and thus

$$(U(Q') - U(Q))\rho(Q, Q') =$$

$$-2 \int exp(iP(Q-Q')) \sum_{k \geq 0} (i/2)^{2k+1} ((2k+1)!)^{-1} (U^{(2k+1)}((Q+Q')/2) \partial_P^{2k+1} W((Q+Q')/2, P)) dP$$

Also note that

$$\rho_{,11}(Q, Q') =$$

$$\partial_Q^2 \int W((Q + Q')/2, P) exp(iP(Q - Q')) dP$$

$$= \int ((1/4) W_{,11}((Q+Q')/2, P)$$

$$-P^2 W((Q+Q')/2, P) + iPW_{,1}((Q+Q')/2, P)) exp(iP(Q-Q')) dP$$

and likewise,

$$\rho_{,22}(Q, Q') = \int ((1/4) W_{,11}((Q+Q')/2, P)$$

$$-P^2 W((Q+Q')/2, P) - iPW_{,1}((Q+Q')/2, P)) exp(iP(Q-Q')) dP$$

3.2 The noiseless quantum Fokker-Planck equation or equivalently, the Liouville-Schrodinger-Von-Neumann-equation in the Wigner domain

$$i\rho'(t) = [(P^2/2m + U(Q)), \rho(t)]$$

is equivalent to

$$i\partial_t(W(t, (Q + Q')/2, P)) =$$

$$(-iP/m)W_{,1}(t, (Q + Q')/2, P)$$

$$+2 \sum_{k \geq 0} (i/2)^{2k+1} ((2k + 1)!)^{-1} (U^{(2k+1)}((Q + Q')/2) \partial_P^{2k+1} W(t, (Q + Q')/2, P))$$

or equivalently,

$$i\partial_t W(t, Q, P) =$$

$$(-iP/m)W_{,1}(t, Q, P) + U'(Q)\partial_P W(t, Q, P) +$$

$$+2 \sum_{k \geq 0} (i/2)^{2k+1} ((2k + 1)!)^{-1} (U^{(2k+1)}(Q) \partial_P^{2k+1} W(t, Q, P))$$

This can be expressed in the form of a classical Liouville equation with quantum correction terms:

$$\partial_t W(t, Q, P) = (-P/m)\partial_Q W(t, Q, P) + U'(Q)\partial_P W(t, Q, P)$$

$$+ \sum_{k \geq 1} ((-1)^k / (2^{2k}(2k + 1)!)) U^{(2k+1)}(Q) \partial_P^{2k+1} W(t, Q, P)$$

The last term is the quantum correction term.

Remark: To see that the last term is indeed a quantum correction term, we have to use general units in which Planck's constant is not set to unity. Thus, we define

$$W(Q, P) = C \int \rho(Q + q/2, Q - q/2).exp(-iPq/h)dq$$

For $\int W(Q, P)dP = \rho(Q, Q)$, we must set

$$2\pi hC = 1$$

Then,

$$\rho(Q + q/2, Q - q/2) = \int W(Q, P)exp(iPq/h)dP$$

Thus, with $P = -ih\partial_Q$, we get

$$[P^2/2m, \rho](Q, Q') = (h^2/2m)(-\rho_{,11}(Q, Q') + \rho_{,22}(Q, Q'))$$

$$= (h^2/2m)(-\partial_Q^2 + \partial_{Q'}^2) \int W((Q + Q')/2, P)exp(iP(Q - Q')/h)dP$$

$$= (-ih/m) \int PW_{,1}((Q + Q')/2, P)exp(iP(Q - Q')/h)dP$$

Further,

$$(U(Q) - U(Q'))\rho(Q, Q') = \sum_{k \geq 0} 2^{-2k}((2k+1)!)^{-1}U^{(2k+1)}((Q+Q')/2)(Q-Q')^{2k+1}\rho(Q, Q')$$

$$= \int \sum_k 2^{-2k}((2k+1)!)^{-1}(ih)^{2k+1}U^{(2k+1)}((Q+Q')/2)\partial_P^{2k+1}W((Q+Q')/2, P)exp(iP(Q-Q')/h)dP$$

and hence its we get from the Schrodinger equation

$$ih\partial_t\rho = [H, \rho] = [P^2/2m, \rho] + [U(Q), \rho],$$

the equation

$$ih\partial_t W(t, Q, P) = (-ih/m)PW_{,1}((Q + Q')/2, P)$$

$$+ihU'(Q)\partial_P W(t, Q, P) + ih\sum_{k \geq 1}(-1)^k 2^{-2k}h^{2k}((2k+1)!)^{-1}U^{(2k+1)}(Q)\partial_P^{2k+1}W(t, Q, P)$$

or equivalently,

$$\partial_t W(t, Q, P) = (-P/m)\partial_Q W(t, Q, P) + U'(Q)\partial_P W(t, Q, P)$$

$$+ \sum_{k \geq 1}(-h^2/4)^k((2k+1)!)^{-1}U^{(2k+1)}(Q)\partial_P^{2k+1}W(t, Q, P)$$

which clearly shows that the quantum correction to the Liouville equation is a power series in h^2 beginning with the first power of h^2. Note that the quantum corrections involve $\partial_P^{2k+1}W(t, Q, P), k = 1, 2, ...$ but a diffusion term of the form $\partial_P^2 W$ is absent. It will be present only when we include the Lindblad terms for it is these terms that describe the effect of a noisy bath on our quantum system.

3.3 Construction of the quantum Fokker-Planck equation for a specific choice of the Lindblad operator

Here,

$$H = P^2/2m + U(Q), L = g(Q)$$

Assume that g is a real function. Then, the Schrodinger equation for the mixed state ρ is given by

$$\partial_t \rho = -i[H, \rho] - (1/2)\theta(\rho)$$

where

$$\theta(\rho) = L^*L\rho + \rho L^*L - 2L\rho L^*$$

Thus, the kernel of this in the position representation is given by

$$\theta(\rho)(Q, Q') = (g(Q)^2 + g(Q')^2 - 2g(Q)g(Q'))\rho(Q, Q')$$

$$= (g(Q) - g(Q'))^2 \rho(Q, Q')$$

More generally, if

$$\theta(\rho) = \sum_k (L_k^* L_k \rho + \rho L_k^* L_k - 2L_k \rho L_k)$$

where

$$L_k = g_k(Q)$$

are real functions, then

$$\theta(\rho)(Q, Q') = F(Q, Q')\rho(Q, Q')$$

where

$$F(Q, Q') = \sum_k (g_k(Q) - g_k(Q'))^2$$

We define

$$G(Q, q) = F(Q + q/2, Q - q/2)$$

or equivalently,

$$F(Q, Q') = G((Q + Q')/2, Q - Q')$$

Then,

$$\theta(\rho)(Q, Q') = G((Q + Q')/2, (Q - Q')\rho(Q, Q') =$$

$$\sum_{n \geq 0} G_n((Q + Q')/2)(Q - Q')^n \int W((Q + Q')/2, P)exp(iP(Q - Q')/h)dP$$

$$= \sum_{n \geq 0} G_n((Q + Q')/2)(ih)^n \int \partial_P^n W((Q + Q')/2, P)exp(iP(Q - Q')/h)dP$$

and this contributes a factor of

$$-(1/2)\sum_{n\geq 0}(ih)^n G_n(Q)\partial_P^n W(Q,P)$$

$$= (-1/2)G_0(Q)W(Q,P) - (ih/2)G_1(Q)\partial_P W(Q,P) + (h^2/2)G_2(Q)\partial_P^2 W(Q,P)$$
$$-(1/2)\sum_{n>2}(ih)^n G_n(Q)\partial_P^n W(Q,P)$$

The term $(h^2/2)G_2(Q)\partial_P^2 W(Q,P)$ is the quantum analogue of the classical diffusion term in the Fokker-Planck equation. Here, we are assuming an expansion

$$G(Q,q) = \sum_n G_n(Q)q^n$$

ie,

$$G_n(Q) = n!^{-1}\partial_q^n G(Q,q)|_{q=0}$$

More generally, consider a Lindblad operator of the form

$$L = \sum_{n\geq 0}g_n(Q)P^n$$

Let us see what the corresponding Lindblad terms contribute to the mixed state evolution. First note that L^*L can be also be expressed in the form

$$L^*L = \sum_{n,m}P^n \bar{g}_n(Q)g_m(Q)P^m = \sum_n h_n(Q)P^n$$

on using the commutation relation

$$[Q,P] = ih$$

Then

$$(L^*L\rho)(Q,Q') = \sum_n h_n(Q)(-ih\partial_Q)^n \rho(Q,Q')$$

and in the Wigner domain, this contributes a term

$$h_n^{m)}(Q)(1/2)^m(ih)^m(-ih)^n(1/2)^k(i/h)^{n-k}\binom{n}{k}\partial_P^m(P^{n-k}\partial_Q^k W(Q,P))$$

Likewise

$$\rho L^*Lf(Q) = \sum_n(-ih)^n\int \rho(Q,Q')h_n(Q')\partial_{Q'}^n f(Q')dQ'$$

$$= \sum_n h^n\int \partial_{Q'}^n(\rho(Q,Q')h_n(Q'))f_n(Q')dQ'$$

yielding

$$(\rho L^* L)(Q, Q') = \sum_n h^n \partial_{Q'}^n (\rho(Q, Q') h_n(Q'))$$

$$= \sum_{n,k} h^n \binom{n}{k} (\partial_{Q'}^k \rho(Q, Q')) \partial_{Q'}^{n-k} h_n(Q')$$

A typical term in this expansion has the form

$$(\partial_{Q'}^k \rho(Q, Q')) f(Q')$$

$$= (\partial_{Q'}^k \rho(Q, Q')) \sum_n n!^{-1} f^n ((Q + Q')/2)((Q - Q')/2)^n$$

Again a typical term in this expansion has the form

$$f((Q + Q')/2)(\partial_{Q'}^k \rho(Q, Q'))(Q - Q')^n$$

and this contributes in the Wigner distribution domain, a term

$$f(Q) \binom{k}{r} (ih)^n (-i/h)^{k-r} \partial_P^n (P^{k-r} \partial_Q^r W(Q, P))$$

3.4 Problems in quantum corrections to classical theories in probability theory and in mechanics with other specific choices of the Lindblad operator

Consider the Lindblad term in the density evolution problem:

$$\theta(\rho) = L^* L \rho + \rho L^* L - 2L\rho L^*)$$

where

$$L = g_0(Q) + g_1(Q)P$$

We have

$$L^* = \bar{g}_0(Q) + P\bar{g}_1(Q)$$

$$L^* L \psi(Q) = \bar{g}_0 + P\bar{g}_1)(g_0 + g_1 P)\psi(Q)$$

$$= |g_0(Q)|^2 \psi(Q) + P|g_1(Q)|^2 P\psi(Q) + \bar{g}_0 g_1 P\psi(Q) + P\bar{g}_1 g_0 \psi(Q)$$

Now, defining

$$|g_1(Q)|^2 = f_1(Q), |g_0(Q)|^2 = f_0(Q),$$

we get

$$P f_1(Q) P\psi(Q) = -\partial_Q f_1(Q)\psi'(Q) = -f_1'(Q)\psi'(Q) - f_1(Q)\psi''(Q)$$

$$P\bar{g}_1(Q)g_0(Q)\psi(Q) = -i(\bar{g}_1 g_0)'(Q)\psi(Q) - \bar{g}_1(Q)g_0 Q)\psi'(Q)$$

Likewise for the other terms. Thus, L^*L has the following kernel in the position representation:

$$L^*L(Q,Q') = -f_1(Q)\delta''(Q-Q') + f_2(Q)\delta'(Q-Q') + f_3(Q)\delta(Q-Q')$$

where $f_1(Q)$ is real positive and f_2, f_3 are complex functions. Thus,

$$L^*L\rho(Q,Q') = -f_1(Q)\partial_Q^2\rho(Q,Q') + f_2(Q)\partial_Q\rho(Q,Q') + f_3(Q)\rho(Q,Q')$$

$$\rho L^*L(Q,Q') = \int \rho(Q,Q'')L^*L(Q'',Q')dQ'' = -\partial_{Q'}^2(\rho(Q,Q')f_1(Q')) -$$

$$\partial_{Q'}(\rho(Q,Q')f_2(Q')) + \rho(Q,Q')f_3(Q')$$

Likewise,

$$L\rho L^*(Q,Q') = g_1(Q)\bar{g}_1(Q')\partial_Q\partial_{Q'}\rho(Q,Q') + T$$

where T contains terms of the form a function of Q, Q' times first order partial derivatives in Q, Q'. Putting all the terms together, we find that tbe Lindblad contribution to the master equation has the form

$$(-1/2)\theta(\rho)(Q,Q') = (f_1(Q)\partial_Q^2 + f_1(Q')\partial_{Q'}^2)\rho(Q,Q') - g_1(Q)\bar{g}_1(Q')\partial_Q\partial_{Q'}\rho(Q,Q') +$$

$$h_1(Q,Q')\partial_Q\rho(Q,Q') + h_2(Q,Q')\partial_{Q'}\rho(Q,Q') + h_3(Q,Q')\rho(Q,Q')$$

3.5   Belavkin filter for the Wigner distribution function

The HPS dynamics is

$$dU(t) = (-(iH + LL^*/2)dt + LdA(t) - L^*dA(t)^*)U(t)$$

where

$$L = aP + bQ, H = P^2/2 + U(Q)$$

The coherent state in which the Belavkin filter is designed to operate is

$$|\phi(u)\rangle = exp(-|u|^2/2)|e(u)\rangle, u \in L^2(\mathbb{R}_+)$$

The measurement process is

$$Y_o(t) = U(t)^*Y_i(t)U(t), Y_i(t) = cA(t) + \bar{c}A(t)^*$$

Then,

$$dY_o(t) = dY_i(t) - j_t(\bar{c}L + cL^*)dt$$

where
$$j_t(X) = U(t)^* X U(t)$$

Note that

$$\bar{c}L + cL^* = \bar{c}(aP + bQ) + c(\bar{a}Q + \bar{b}P) = 2Re(\bar{a}c)Q + 2Re(\bar{b}c)P$$

Thus, our measurement model corresponds to measuring the observable $2Re(\bar{a}c)Q(t) + 2Re(\bar{b}c)P(t)$ plus white Gaussian noise. Let

$$\pi_t(X) = \mathbb{E}(j_t(X)|\eta_o(t)), \eta_o(t) = \sigma(Y_o(s) : s \leq t)$$

Let
$$d\pi_t(X) = F_t(X)dt + G_t(X)dY_o(t)$$

with
$$F_t(X), G_t(X) \in \eta_o(t)$$

Let
$$dC(t) = f(t)C(t)dY_o(t), C(0) = 1$$

The orthogonality principle gives

$$\mathbb{E}[(j_t(X) - \pi_t(X))C(t)] = 0$$

and hence by quantum Ito's formula and the arbitrariness of the complex valued function $f(t)$, we have

$$\mathbb{E}[(dj_t(X) - d\pi_t(X))|\eta_o(t)] = 0,$$

$$\mathbb{E}[(j_t(X) - \pi_t(X))dY_o(t)|\eta_o(t)] + \mathbb{E}[(dj_t(X) - d\pi_t(X))dY_o(t)|\eta_o(t)] = 0$$

Note that

$$dj_t(X) = j_t(\theta_0(X))dt + j_t(\theta_1(X))dA(t) + j_t(\theta_2(X))dA(t)^*$$

where

$$\theta_0(X) = i[H, X] - (1/2)(LL^*X + XLL^* - 2LXL^*),$$

$$\theta_1(X) = -LX + XL = [X, L], \theta_2(X) = L^*X - XL^* = [L^*, X]$$

Then the above conditions give

$$\pi_t(\theta_0(X)) + \pi_t(\theta_1(X))u(t) + \pi_t(\theta_2(X))\bar{u}(t) =$$

$$F_t(X) + G_t(X)(cu(t) + \bar{c}\bar{u}(t) - \pi_t(\bar{c}L + cL^*))$$

and

$$-\pi_t(X(\bar{c}L + cL^*)) + \pi_t(X)\pi_t(\bar{c}L + cL^*) + \pi_t(\theta_1(X))\bar{c}$$
$$-G_t(X)|c|^2 = 0$$

This second equation simplifies to

$$G_t(X) = |c|^{-2}(-\pi_t(cXL^* + \bar{c}LX) + \pi_t(\bar{c}L + cL^*)\pi_t(X))$$

Then,
$$d\pi_t(X) = (\pi_t(\theta_0(X)) + \pi_t(\theta_1(X))u(t) + \pi_t(\theta_2(X))\bar{u}(t))dt$$
$$+G_t(X)(dY_o(t) + \pi_t(\bar{c}L + cL^* - cu(t) - \bar{c}\bar{u}(t))dt)$$

In the special case when $c = -1$, we get
$$d\pi_t(X) = (\pi_t(\theta_0(X)) + \pi_t(\theta_1(X))u(t) + \pi_t(\theta_2(X))\bar{u}(t))dt$$
$$+(\pi_t(XL^* + LX) - \pi_t(L + L^*)\pi_t(X))(dY_o(t) - (\pi_t(L + L^*) - 2Re(u(t)))dt)$$

Consider the simplified case when $u = 0$, ie, filtering is carried out in the vacuum coherent state. Then, the above Belavkin filter simplifies to

$$d\pi_t(X) = \pi_t(\theta_0(X))dt + (\pi_t(XL^* + LX) - \pi_t(L + L^*)\pi_t(X))(dY_o(t) - \pi_t(L + L^*)dt)$$

Equivalently in the conditional density domain, the dual of the above equation gives the Belavkin filter for the conditional density

$$\rho'(t) = \theta_0^*(\rho(t))dt + (L^*\rho(t) + \rho(t)L - Tr(\rho(t)(L + L^*))\rho(t))(dY_o(t) - Tr(\rho(t)(L + L^*))dt)$$

where
$$\theta_0^*(\rho) = -i[H, \rho] - (1/2)(LL^*\rho + \rho LL^* - 2L^*\rho L)$$

Now let us take
$$L = aQ + bP, a, b \in \mathbb{C}$$
and
$$H = P^2/2 + U(Q)$$

and translate this Belavkin equation to the Wigner-distribution domain and then compare the resulting differential equation for W with the Kushner-Kallianpur filter for the classical conditional density. We write

$$\rho(t, Q, Q') = \int W(t, (Q + Q')/2, P)exp(iP(Q - Q'))dP$$

Note that the Kushner-Kallianpur filter for the corresponding classical probabilistic problem

$$dQ = Pdt, dP = -U'(Q)dt - \gamma Pdt + \sigma dB(t),$$

$$dY(t) = (2Re(a)Q(t) + 2Re(b)P(t))dt + \sigma dV(t)$$

where B, V are independent identical Brownian motion processes is given by

$$dp(t, Q, P) = L^*p(t, Q, P) + ((\alpha Q + \beta P)p(t, Q, P)$$

$$-(\int(\alpha Q + \beta P)p(t, Q, P)dQdP)p(t, Q, P))(dY(t) - dt\int(\alpha Q + \beta P)p(t, Q, P)dQdP)$$

where
$$\alpha = 2Re(a), \beta = 2Re(b)$$

where

$$L^*p = -P\partial_Q p + U'(Q)\partial_P p + \gamma.\partial_P(Pf) + (\sigma^2/2)\partial_P^2 p$$

Let us see how this Kushner-Kallianpur equation translates in the quantum case using the Wigner distribution W in place of p. As noted earlier, the equation

$$\rho'(t) = -i[H, \rho(t)]$$

is equivalent to

$$\partial_t W(t, Q, P) = (-P/m)\partial_Q W(t, Q, P) + U'(Q)\partial_P W(t, Q, P)$$

$$+ \sum_{k \geq 1} (-h^2/4)^k ((2k+1)!)^{-1} U^{(2k+1)}(Q)\partial_P^{2k+1} W(t, Q, P) \; - - - (a)$$

while the Lindblad correction term to (a) is $(-1/2)$ times

$$LL^*\rho + \rho LL^* - 2L^*\rho L$$

$$= (aQ + bP)(\bar{a}Q + \bar{b}P)\rho + \rho(aQ + bP)(\bar{a}Q + \bar{b}P) - 2(\bar{a}Q + \bar{b}P)\rho(aQ + bP)$$

which in the position representation has a kernel of the form

$$-|b|^2(\partial_Q^2 + \partial_{Q'}^2 + 2\partial_Q\partial_{Q'}) - |a|^2(Q - Q')^2$$

$$+(c_1 Q + c_2 Q' + c_3)\partial_Q + (c_4 Q + c_5 Q' + c_6)\partial_{Q'}]\rho(t, Q, Q')$$

This gives a Lindblad correction to (a) of the form

$$(\sigma_1^2/2)\partial_Q^2 W(t, Q, P) + (\sigma_2^2/2)\partial_P^2 W(t, Q, P)$$

$$+d_1\partial_Q\partial_P W(t, Q, P) + (d_3 Q + d_4 P + d_5)\partial_Q W(t, Q, P)$$

Thus, the master/Lindblad equation (in the absence of measurements) has the form

$$\partial_t W(t, Q, P) = (-P/m)\partial_Q W(t, Q, P) + U'(Q)\partial_P W(t, Q, P)$$

$$+ \sum_{k \geq 1} (-h^2/4)^k ((2k+1)!)^{-1} U^{(2k+1)}(Q)\partial_P^{2k+1} W(t, Q, P)$$

$$+(\sigma_1^2/2)\partial_Q^2 W(t, Q, P) + (\sigma_2^2/2)\partial_P^2 W(t, Q, P)$$

$$+d_1\partial_Q\partial_P W(t, Q, P) + (d_3 Q + d_4 P + d_5)\partial_Q W(t, Q, P)$$

Remark: Suppose $b = 0$ and a is replaced by a/h. Then, taking into account the fact that in general units, the factor $exp(iP(Q - Q'))$ must be replaced by $exp(iP(Q - Q'))/h$ in the expression for $\rho(t, Q, Q')$ in terms of $W(t, Q, P)$, we get the Lindblad correction as

$$(|a|^2/2)(Q - Q')^2\rho(t, Q, Q')$$

or equivalently in the Wigner-distribution domain as

$$(|a|^2/2)\partial_P^2 W(t, Q, P)$$

and the master equation simplifies to

$$\partial_t W(t,Q,P) = (-P/m)\partial_Q W(t,Q,P) + U'(Q)\partial_P W(t,Q,P) + (|a|^2/2)\partial_P^2 W(t,Q,P)$$

$$+ \sum_{k \geq 1} (-h^2/4)^k ((2k+1)!)^{-1} U^{(2k+1)}(Q)\partial_P^{2k+1} W(t,Q,P)$$

This equation clearly displays the classical terms and the quantum correction terms.

Exercise:Now express the terms $L^*\rho(t), \rho(t)L^*$ and $Tr(\rho(t)(L+L^*))$ in terms of the Wigner distribution for $\rho(t)$ assuming $L = aQ + bP$ and complete the formulation of the Belavkin filter in terms of the Wigner distribution. Compare with the corresponding classical Kushner-Kallianpur filter for the conditional density of $(Q(t), P(t))$ given noisy measurements of $aQ(s) + bP(s), s \leq t$ by identifying the precise form of the quantum correction terms.

3.6 Superstring coupled to gravitino ensures local supersymmetry

The action is

$$S = \int e[(1/2)\partial_\alpha X^\mu \partial^\alpha X_\mu + \bar\psi^\mu \rho^\alpha \partial_\alpha \psi_\mu + \bar\chi_\alpha \partial_\beta X^\mu \rho^\beta \rho^\alpha \psi_\mu + \bar\psi^\mu \psi_\mu \bar\chi_\beta \rho^\beta \rho^\alpha \chi_\alpha] d^2\sigma$$

where the string world-sheet metric is $h_{\alpha\beta}$ and its dyad representation is

$$h_{\alpha\beta}(\tau,\sigma) = e_\alpha^a e_\beta^a = \eta_{ab} e_\alpha^a e_\beta^b$$

and

$$e = \sqrt{h} = e = deta(e_\alpha^a)$$

The local supersymmetry transformation that leave this action invariant are

$$\delta X^\mu = \bar\epsilon \psi^\mu,$$

$$\delta\psi^\mu = \rho^\alpha \epsilon(\partial_\alpha X^\mu - \bar\chi_\alpha \psi^\mu),$$

$$\delta\chi_\alpha = \nabla_\alpha \epsilon,$$

$$\delta e_\alpha^a = c\bar\epsilon \rho^a \chi_\alpha$$

Problem: Develop a quantum filtering theory for the above superstring action coupled to gravitino in the Wigner domain by expressing the superstring action in terms of Bosonic and Fermionic creation and annihilation operators and formulating the Belavkin filter equations followed by transformation to the Wigner domain.

Chapter 4

Undergraduate and Postgraduate Courses in Electronics, Communication and Signal Processings

1. Basic electronic circuits, digital and statistical signal processing, digital and analog communication, device physics, nonlinear system theory, digital image processing including maximum entropy histogram equalization as well as modern areas of signal processing like quantum computation, quantum information and commmunication and quantum filtering.

2. Matrix computations for implementing the quantum Belavkin filter and corresponding quantum control algorithms based on the Hudson-Parthasarathy quantum stochastic calculus.

3. How quantum filtering can be applied to estimating the spin of an electron from noisy measurements

4. Reduction of quantum noise using Luc-Bouten's method of quantum control.

5. Quantum entropy generated in a noisy quantum system by the bath and how filtering can be used to reduce this entropy.

6. Implementing the quantum Schrodinger channel which involves using a quantum receiver based on Schrodinger's equation to detect weak electromagnetic signals using the principle of quantum measurement and state collapse.

Chapter 5

Quantization of Classical Field Theories, Examples

5.1 Quantization of fluid dynamics in a curved space-time background using Lagrange multiplier functions

First consider the quantization of non-relativistic fluid dynamics. The Lagrangian density is

$$L(v, v_{,0}, v_{,j}, u, u_{,0}, p, w) = u_i(v_{i,0} - F_i(v, v_{,j}, p_{,j})) + \epsilon v_{i,0}^2/2 + \mu u_{i,0}^2/2$$

$$+w.v_{i,i}$$

where u_i, w are Lagrange multiplier fields, ϵ, μ are small numbers converging to zero and

$$F_i(v, v_{,j}, p) = -v_j v_{i,j} + \eta v_{i,jj} - p_{,i}$$

The Euler Lagrange equations obtained from

$$\delta_{u_i} \int L d^4 x = 0$$

are

$$v_{i,0} - F_i(v, v_{,j}, p_{,j}) - \mu u_{i,00} = 0 - - - (1)$$

which is the Navier-Stokes equation with a correction term $\mu u_{i,00}$.

Remark: The fact that ϵ is small in the Lagrangian means that we do not allow the acceleration $v_{i,0}$ to get too large. Indeed, if it is too large, then the forces will be too large and the fluid dynamical approximation would break down. The other Euler-Lagrange equations

$$\delta_w \int L d^4 x = 0$$

29

gives

$$v_{i,i} = 0 - -(2)$$

which is the incompressibility equation. The other Euler-Lagrange equations are

$$\delta_{v_i} \int L d^4 x = 0, \delta_p \int L d^4 x = 0$$

which give

$$-u_{i,0} + u_j v_{j,i} + (u_i v_j)_{,j} - \epsilon v_{i,00} - w_{,i} = 0$$

which in view of the incompressibility equation, simplifies to

$$-u_{i,0} + u_j v_{j,i} + u_{i,j} v_j - \epsilon v_{i,00} - w_{,i} = 0 - -(3)$$

and

$$u_{i,i} = 0 - -(4)$$

equations (1)-(4) are our basic equations totally $3 + 1 + 3 + 1 = 8$ in number for our eight fields v_i, p, u_i, w. Now let us calculate the Hamiltonian density. To get non-trivial momentum fields corresponding to p and w, we include more regulatory terms in the Lagrangian which say that their rates of change cannot be too large:

$$L(v, v_{,0}, v_{,j}, u, u_{,0}, , p_{,0}, w_{,0}, p, w) = u_i(v_{i,0} - F_i(v, v_{,j}, p_{,j})) + \epsilon_1 v_{i,0}^2/2 + \mu_1 u_{i,0}^2/2$$

$$+w.v_{i,i} + \epsilon_2 p_{,0}^2/2 + \mu_2 w_{,0}^2/2$$

and then the above Euler-Lagrange equations get modified to

$$v_{i,0} - F_i(v, v_{,j}, p_{,j}) - \mu u_{i,00} = 0 - - - (1')$$

$$v_{i,i} - \mu_2 w_{,00} = 0 - -(2')$$

$$-u_{i,0} + u_j v_{j,i} + u_{i,j} v_j - \epsilon v_{i,00} - w_{,i} = 0 - -(3')$$

$$u_{i,i} - \epsilon_2 p_{,00} = 0 - -(4')$$

The canonical momentum fields are

$$P_{v_i} = \partial L/\partial v_{i,0} = u_i + \epsilon_1 v_{i,0},$$

$$P_{u_i} = \partial L/\partial u_{i,0} = \mu_1 u_{i,0},$$

$$P_p = \partial L/\partial p_{,0} = \epsilon_2 p_{,0},$$

$$P_w = \partial L/\partial w_{,0} = \mu_2 w_{,0}$$

Then the Hamiltonian density is

$$H = P_{v_i} v_{i,0} + P_{u_i} u_{i,0} + P_p p_{,0} + P_w w_{,0} - L$$

5.2 d-dimensional harmonic oscillator with electric field forcing

The Hamiltonian is

$$H(t) = \sum_{k=1}^{d} \omega(k)a(k)^*a(k) - \sum_{k=1}^{d} E_k(t)(a(k) + a(k)^*)$$

where

$$[a(k), a(m)^*] = \delta(k, m), [a(k), a(m)] = [a(k)^*, a(m)^*] = 0$$

Let

$$b(k,t) = a(k) - E_k(t)/\omega(k), b(k,t)^* = a(k)^* - E_k(t)/\omega(k)$$

Then

$$H(t) = \sum_{k} \omega(k)b(k,t)^*b(k,t) - c(t)$$

where

$$c(t) = \sum_{k} E_k(t)^2/\omega(k)$$

is a c-number function of time. We have

$$[b(k,t), b(m,t)^*] = \delta(k,m), [b(k,t), b(m,t)] = [b(k,t)^*, b(m,t)^*] = 0$$

Let

$$\mathbf{n} = (n_1, ..., n_d) \in \mathbb{Z}_+^d = \{0, 1, ...,\}^d$$

The state $|\mathbf{0} >_t$ is defined by

$$b(k,t)|\mathbf{0} >_t = 0, k = 1, 2, ..., d$$

This gives with

$$a(k) = (q(k) + ip(k))/\sqrt{2}, a(k)^* = (q(k) - ip(k))/\sqrt{2}, [q(k), p(m)] = i\delta(k, m),$$

$$[q(k), q(m)] = 0 = [p(k), p(m)],$$

the equation

$$(q(k) + \partial/\partial q(k) - \sqrt{2}E_k(t)/\omega(k))|\mathbf{0} >_t = 0$$

so that

$$< \mathbf{q}|\mathbf{0} >_t = C(t).exp(-\sum_{k}(q(k)^2/2 + \sqrt{2}E_k(t)q(k)/\omega(k)))$$

where

$$C(t) = \pi^{-d/4}.exp(-\sum_{k} E_k(t)^2/\omega(k)^2)$$

Let

$$|\mathbf{n} >_t = (\Pi_k b(k,t)^{*n_k}/\sqrt{n_k!})|\mathbf{0} >_t$$

$< \mathbf{q}|\mathbf{n}>_t, \mathbf{n} \in \mathbb{Z}_+$ form an onb for $L^2(\mathbb{R}^d)$. Now expand the wave function as

$$|\psi(t)> = \sum_{\mathbf{n}} c(\mathbf{n}, t)|\mathbf{n}>_t$$

Note that

$$H(t)|\mathbf{n}>_t = (\sum_k n_k \omega(k) - c(t))|\mathbf{n}>_t$$

$$i\partial_t|\psi(t)> = H(t)|\psi(t)>$$

$$\partial_t|\mathbf{n}>_t = \sum_m [\Pi_{k=m}b(k,t)^{*n_k}/\sqrt{n_k!}]((n_m-1)!)^{-1/2}$$

$$\sqrt{n_m}b(m,t)^{*n_m-1}(-E'_m(t)/\omega(m))|0>_t$$

$$+(\Pi_k b(k,t)^{*n_k}\sqrt{n_k!})\partial_t|0>_t$$

$$= \sum_m (-E'_m(t)/\omega(m))\sqrt{n_m}|\mathbf{n} - \mathbf{e}_m>_t$$

$$+(\sum_m \sqrt{2}E'_m(t)/\omega(m))(\Pi_k b(k,t)^{*n_k}/\sqrt{n_k!})q(m)|0>_t$$

$$+(C'(t)/C(t))|\mathbf{n}>_t$$

Now,

$$q(m)|0>_t = ((a(m)+a(m)^*)/\sqrt{2})|0>_t$$

$$= ((b(m,t)+b(m,t)^*)/\sqrt{2}) + \sqrt{2}E_m(t)/\omega(m))|0>_t$$

$$= (\sqrt{2}E_m(t)/\omega(m) + b(m,t)^*/\sqrt{2})|0>_t$$

Thus,

$$(\Pi_k b(k,t)^{*n_k}/\sqrt{n_k!})q(m)|0>_t$$

$$= (\sqrt{2}E_m(t)/\omega(m))|\mathbf{n}>_t + 2^{-1/2}\sqrt{n_m+1}|\mathbf{n}+\mathbf{e}_m>_t$$

Therefore,

$$\partial_t|\mathbf{n}>_t =$$

$$= \sum_m (-E'_m(t)/\omega(m))\sqrt{n_m}|\mathbf{n} - \mathbf{e}_m>_t$$

$$+\sum_m [(\sqrt{2}E'_m(t)/\omega(m))((\sqrt{2}E_m(t)/\omega(m))|\mathbf{n}>_t + 2^{-1/2}\sqrt{n_m+1}|\mathbf{n}+\mathbf{e}_m>_t)]$$

$$+(C'(t)/C(t))|\mathbf{n}>_t$$

where $\mathbf{e}_m$ is the $d \times 1$ vector with a one at the mth position and a zero at all the other positions. Thus,

$$\partial_t|\psi(t)> = \sum_{\mathbf{n}}[\partial_t c(\mathbf{n},t))|\mathbf{n}>_t + c(\mathbf{n},t)\partial_t|\mathbf{n}>_t]$$

$$= \sum_{\mathbf{n}} \partial_t c(\mathbf{n},t)|\mathbf{n}>_t - \sum_{\mathbf{n},m}[c(\mathbf{n},t)(E'_m(t)/\omega(m))\sqrt{n_m}|\mathbf{n} - \mathbf{e}_m>_t]$$

$$+\sum_{n,m}[c(\mathbf{n},t)(\sqrt{2}E'_m(t)/\omega(m))((\sqrt{2}E_m(t)/\omega(m))|\mathbf{n}>_t +2^{-1/2}\sqrt{n_m+1}|\mathbf{n}+\mathbf{e}_m>_t)]$$

$$+(C'(t)/C(t))\sum_{\mathbf{n}}c(\mathbf{n},t)|\mathbf{n}>_t$$

or equivalently,

$$<\mathbf{n}|_t\partial_t\psi(t)>=$$

$$\partial_t c(\mathbf{n},t)-\sum_m(E'_m(t)/\omega(m))\sqrt{n_m}c(\mathbf{n}+\mathbf{e}_m,t)$$

$$+\sum_m[(\sqrt{2}E'_m(t)/\omega(m))((\sqrt{2}E_m(t)/\omega(m))c(\mathbf{n},t)+\sqrt{n_m+1}2^{-1/2}c(\mathbf{n}-\mathbf{e}_m,t))]$$

$$+(C'(t)/C(t))c(\mathbf{n},t)$$

So the Schrodinger equation

$$<\mathbf{n}|_t i\partial_t|\psi(t)>=<\mathbf{n}|_t H(t)|\psi(t)>$$

is equivalent to the infinite sequence of linear differential equations

$$i\partial_t c(\mathbf{n},t)-i\sum_m(E'_m(t)/\omega(m))\sqrt{n_m}c(\mathbf{n}+\mathbf{e}_m,t)$$

$$+i\sum_m[(\sqrt{2}E'_m(t)/\omega(m))((\sqrt{2}E_m(t)/\omega(m))c(\mathbf{n},t)+\sqrt{n_m+1}2^{-1/2}c(\mathbf{n}-\mathbf{e}_m,t))]$$

$$+(iC'(t)/C(t))c(\mathbf{n},t)$$

$$=(\sum_m\omega(m)n_m-c(t))c(\mathbf{n},t)$$

5.3 A problem:Design a quantum neural network based on matching the diagonal slice of the density operator to a given probability density function

5.4 Quantum filtering for the gravitational field interacting with the electromagnetic field

Let $\mathbf{Q},\mathbf{P}$ denote the position and momentum observables of the gravitational field with Lagrangian $L_1(\mathbf{Q},\mathbf{P})$, namely the Einstein-Hilbert action in the ADM formalism. Let $L_2(\mathbf{Q},\mathbf{P},A_\mu,A_{\mu,\nu})$ denote the Maxwell Lagrangian $g^{\mu\alpha}g^{\nu\beta}\sqrt{-g}F_{\mu\nu}F_{\alpha\beta}$ with

$$F_{\mu\nu}=A_{\nu,\mu}-A_{\mu,\nu}$$

We can choose a coordinate system so that $h_{\mu 0} = 0$ where

$$g_{\mu\nu}(x) = \eta_{\mu\nu} + h_{\mu\nu}(x)$$

Note that in the ADM system, the position variables are $\tilde{g}_{ab}, 1 \leq a \leq b \leq 3$ and N, N^a where the tilde above the metric corresponds to defining the metric in a coordinate system x^μ whose constant time surfaces $x^0 = t$ correspond to a hypersurface Σ_t embedded in $\mathbb{R}^4$. The unit normal on this surface is denoted by n^μ. Events in $\mathbb{R}^4$ are specified by the space-time coordinates X^μ. Thus, we have the orthogonal decomposition

$$X^\mu_{,0} = T^\mu = N^\mu + N n^\mu$$

where

$$N^\mu = N^a X^\mu_{,a}$$

is a spatial vector $N^a, a = 1, 2, 3$ are chosen to satisfy the orthogonality relation

$$g_{\mu\nu} X^\mu_{,a} n^\nu = 0$$

or equivalently,

$$g_{\mu\nu} X^\mu_{,a}(T^\nu - N^\nu) = 0$$

or equivalently

$$g_{\mu\nu} X^\mu_{,a}(T^\nu - N^b X^\nu_{,b}) = 0$$

or equivalently,

$$\tilde{g}_{a0} = \tilde{g}_{ab} N^b$$

Define

$$q_{ab} = \tilde{g}_{ab}$$

Also define

$$K'_{\mu\nu} = \nabla_\mu n_\nu, K_{\mu\nu} = q^{\mu'}_\mu q^{\nu'}_\nu K'_{\mu'\nu'}$$

Since n_μ is the normal to a hypersurface, we can write

$$n_\mu = f g_{,\mu}$$

for two scalar functions f, g. Then,

$$\nabla_\mu n_\nu - \nabla_\mu n_\nu =$$

$$n_{\nu,\mu} - n_{\mu,\nu} = f_{,\mu} g_{,\nu} - f_{,\nu} g_{,\mu}$$

$$= (log f)_{,\mu} n_\nu - (log f)_{,\nu} n_\mu$$

Thus, since

$$q^{\nu'}_\nu n_{\nu'} = 0$$

we get

$$K_{\mu\nu} = K_{\nu\mu}$$

and hence if we write

$$K_{ab} = X^\mu_{,a} X^\nu_{,b} \nabla_\mu n_\nu$$

then we have the symmetry:

$$K_{ab} = K_{ba}$$

It is clear that the Einstein Hilbert action in 4 dimensions can be expressed as the sum of the Einstein Hilbert action in three spatial dimensions corresponding to the spatial metric q_{ab} and its spatial derivatives $q_{ab,c}$ plus a quadratic function of the K_{ab} (Ref:Thomas Thiemann, Modern canonical quantum general relativity). Further K_{ab} can be expressed as a linear function of the $q_{ab,0}$ and hence the canonical momentum corresponding to this action $\partial L/\partial q_{ab,0} = P_{ab}$ will again be linear in the $q_{ab,0}$. Further the time derivatives of the position fields N^a, N do not appear in the Einstein-Hilbert action. We have the decomposition

$$g^{\mu\nu} = q^{\mu\nu} + n^\mu n^\nu$$

where

$$q^{\mu\nu} = q_{ab} X^\mu_{,a} X^\nu_{,b}$$

is a purely spatial tensor. This identity can be verified by using the transformation formula

$$g^{\mu\nu} = \tilde{g}^{\alpha\beta} X^\mu_{,\alpha} X^{\nu,\beta} =$$

$$= \tilde{g}^{00} T^\mu T^\nu + \tilde{g}^{0a} (T^\mu X^\nu_{,a} + T^\nu X^\mu_{,a})$$

$$+ \tilde{g}^{ab} X^\mu_{,a} X^\nu_{,b}$$

with the substitution

$$T^\mu = N^a X^\mu_{,a} + N n^\mu$$

The final form of the ADM action will contain N, N^a as Lagrange multipliers. This can be seen from the fact that

$$g = det((g_{\mu\nu}))$$

Now

$$\tilde{g}_{\alpha\beta} = g_{\mu\nu} X^\mu_{,\alpha} X^\nu_{,\beta}$$

implies

$$det((g_{\mu\nu})).det((X^\mu_{,\alpha}))^2 = det((\tilde{g}_{\alpha\beta}))$$

and using

$$\tilde{g}_{ab} = q_{ab}, \tilde{g}_{a0} = q_{ab} N^b$$

and further,

$$g_{\mu\nu} (T^\mu - N^\mu) T^\nu = N^2$$

gives

$$N^2 = \tilde{g}_{00} - \tilde{g}_{a0} N^a$$

so that

$$\tilde{g}_{00} = N^2 + q_{ab} N^a N^b$$

we can obtain $det((\tilde{g}_{\alpha\beta}))$ in terms of q, N^a, N. Writing $\mathbf{Q}$ for $((q_{ab}))$, $\mathbf{N}$ and (N^a), we get

$$det(\tilde{g}_{\alpha\beta})) = \tilde{g} =$$

$$det \begin{pmatrix} N^2 + \mathbf{N}^T\mathbf{Q}\mathbf{N} & \mathbf{N}^T\mathbf{Q} \\ \mathbf{Q}\mathbf{N} & \mathbf{Q} \end{pmatrix}$$

$$= det \begin{pmatrix} N^2 & \mathbf{N}^T\mathbf{Q} \\ \mathbf{0} & \mathbf{Q} \end{pmatrix}$$

$$= N^2 q$$

where

$$q = det(\mathbf{Q})$$

Thus, if $\mathcal{L}$ is the Einstein-Hilbert Lagrangian density R expressed in terms of $q_{ab}, q_{ab,0}, N^a, N$ and P^{ab} the canonical momentum conjugate to q_{ab}, then the Einstein-Hilbert Hamiltonian density is

$$\mathcal{H} = (P^{ab}q_{ab,0} - \mathcal{L})\sqrt{-q}N$$

The other constraint operator involving N^a will appear in this Hamiltonian, once we impose the diffeomorphism constraint. Taking all this into account, it is easy to see that the Einstein-Hilbert Hamiltonian can be expressed as

$$H = \int (N^a H_a + N H_0)d^4x$$

. where H_a, H_0 are only functions of $q_{rs}, q_{rs,m}, P^{rs}$ with H_0 being quadratic in the P^{rs} and H_a linear in the P^{rs}. For further details, the reader can consult the book "Modern Canonical Quantum General Relativity" by Thomas Thiemann, Cambridge University Press. To formulate quantum filtering theory, we must first observe that for Hamiltonian systems with constraints, the Dirac bracket replaces the Lie bracket. This is discussed in what follows.

5.5 Quantum Belavkin filtering for constrained Hamiltonians

Let $H(Q, P)$ be the Hamiltonian and let the constraints be

$$\chi_r(Q, P) = 0, r = 1, 2, ..., m$$

We form the Lie brackets

$$[\chi_r(Q, P), \chi_s(Q, P)] = C_{rs}(Q, P)$$

and then define the Dirac brackets between two observables X, Y as

$$[X, Y]_D = [X, Y] - [X, \chi_r]C^{rs}(Q, P)[\chi_s, Y]$$

where
$$((C^{rs}(Q,P))) = ((C_{rs}(Q,P)))^{-1}$$

and the summation over the repeated indices r,s is implied. Then, we have

$$[X, \chi_m]_D = [X, \chi_m] = [X, \chi_r]C^{rs}[\chi_s, \chi_m]$$

$$= [X, \chi_m] - [X, \chi_r]C^{rs}C_{sm} =$$

$$= [X, \chi_m] - [X, \chi_r]\delta^r_m = 0$$

which agrees with the prescription that since the constraint functions vanish, their commutators with any observable should also vanish and in particular, their commutators with the Hamiltonian should vanish, ie, their time rate of change should vanish. Now the Heisenberg equations of motion of this constrained system are

$$dX(t)/dt = i[H, X(t)]_D$$

Let $\rho(t)$ denote the density matrix. Then if the Schrodinger and Heisenberg pictures do agree for the constrained system, we should have

$$Tr(\rho'(t)X(0)) = Tr(\rho(0)X'(t))$$

$$= iTr(\rho(0)[H, X(t)]_D)$$

Now,
$$Tr(\rho(0)[H, X]_D) = Tr(\rho(0)[H, X])$$

$$-Tr(\rho(0)[H, \chi_r]C^{rs}[\chi_s, X])$$

$$= Tr([\rho(0), H]X))$$

$$-Tr([\rho(0)[H, \chi_r]C^{rs}, \chi_s]X)$$

Thus,the Schrodinger equation for the state of the constrained system is expressible as

$$\rho'(t) = i[\rho(t), H] - i[\rho(t)[H, \chi_r]C^{rs}, \chi_s]$$

$$= i[\rho(t), H] - i[\rho(t), \chi_s][H, \chi_r]C^{rs} - i\rho(t)[[H, \chi_r]C^{rs}, \chi_s]$$

Example: The electromagnetic field. Here, the canonical position fields are $A_r, r = 1, 2, 3$. A_0 is a matter field since if we adopt the Coulomb gauge, $divA = 0$, ie, $A_{r,r} = 0$ and then the Maxwell equations with this gauge condition give

$$\nabla^2 A_0 = -\mu J_0$$

showing that A_0 is indeed a matter field. $divA = 0$ is a constraint. Further, the canonical momentum corresponding to the position field A_r is obtained from the Lagrangian $L = (-1/4)F_{\mu\nu}F^{\mu\nu}$ as

$$\Pi^r = \partial L/\partial A_{r,0} = \partial L/\partial F_{0r}$$

$$-F^{0r} = F_{0r} = E_r, r = 1, 2, 3$$

where E is the electric field. The Maxwell equation $div E = J_0$ then gives the other constraint equation

$$div \Pi - J_0 = 0$$

or equivalently,

$$\Pi^r_{,r} = J_0$$

We therefore have the two constraint functions

$$\chi_1 = A_{r,r} = div A, \chi_2 = \Pi^r_{,r} - J_0 = div \Pi - J_0$$

We evaluate their equal commutators:

$$C_{12}(r, r') = [\chi_1(t, r), \chi_2(t, r')] = [A_{r,r}(t, r)], \Pi^s_{,s}(t, r')] =$$

$$\partial_r \partial'_s [A_r(t, r), \Pi^s(t, r')] =$$

$$\partial_r \partial'_s i \delta^s_r \delta^3(r - r')$$

$$= -i \delta^s_r \partial_r \partial_s \delta^3(r - r')$$

$$-i \nabla^2 \delta(r - r')$$

The inverse kernel of this commutator is

$$\frac{i}{4\pi |r - r'|}$$

Note that

$$C_{21}(r, r') = [\chi_2(t, r), \chi_1(t, r')] =$$

$$-C_{12}(r', r)$$

The inverse kernel of

$$\begin{pmatrix} 0 & C_{12}(r, r') \\ C_{21}(r, r') & 0 \end{pmatrix}$$

is thus

$$\begin{pmatrix} 0 & -i/4\pi |r - r'| \\ i/4\pi |r - r'| & 0 \end{pmatrix}$$

$$= \begin{pmatrix} 0 & C^{12}(r, r') \\ C^{21}(r, r') & 0 \end{pmatrix}$$

Thus, the canonical equal time commutation relations get modified when we use the Dirac bracket in place of the Lie bracket to

$$[A_r(r), \Pi^s(r')]_D = i \delta^s_r \delta^3(r - r') - \int [A_r(r), \chi_m(r_1)] C^{ml}(r_1, r_2)[\chi_l(r_2), \Pi^s(r')] d^3 r_1 d^3 r_2$$

where the sum is over $m, l = 1, 2$ with $m \neq l$. Now,

$$[A_r(r), \chi_1(r_1)] = 0, [A_r(r), \chi_2(r_1)] = [A_r(r), \Pi^s_{,s}(r_1)]$$

$$= -i\partial_r \delta^3(r - r_1),$$

$$[\chi_2(r_2), \Pi^s(r')] = 0, [\chi_1(r_2), \Pi^s(r')] =$$

$$[A_{r,r}(r_2), \Pi^s(r')] = i\partial_s \delta^3(r_2 - r')$$

and combining these results, we get

$$[A_r(r), \Pi^s(r')]_D = i\delta_r^s \delta^3(r - r')$$

$$- \int (-i\partial_r \delta^3(r - r_1)) C^{21}(r_1, r_2)(i\partial_s \delta^3(r_2 - r')) d^3 r_1 d^3 r_2$$

$$= i\delta_r^s \delta^3(r - r') + (\frac{\partial^2}{\partial x_r \partial x'_s}) C^{21}(r, r')$$

$$= i\delta_r^s \delta^3(r - r') + (\frac{\partial^2}{\partial x_r \partial x'_s})(i/(4\pi|r - r'|))$$

Remark on Poisson brackets and Dirac brackets in classical mechanics: Let

$$[u, v] = \sum_{k=1}^{n} (\frac{\partial u(Q, P)}{\partial Q_i} \frac{\partial v}{\partial P_i} - \frac{\partial u(Q, P)}{\partial P_i} \frac{\partial v(Q, P)}{\partial Q_i})$$

$\{u, v\}$ is called the Poisson bracket between the observables u, v in classical mechanics. Now suppose that $Q_k, P_k, k = m + 1, ..., n$ are zero and that these form the $2(n - m)$ constraint equations. Then, since

$$[Q_k, P_m] = \delta_{km}, [Q_k, Q_m] = [P_k, P_m] = 0$$

it follows that the Dirac bracket between u, v is given by

$$[u, v]_D = [u, v] - \sum_{r,s}([u, Q_r](-\delta_{rs})[P_s, v] + [u, P_r]\delta_{rs}[Q_s, v])$$

$$= [u, v] - \sum_{r=m+1}^{n} (\frac{\partial u}{\partial Q_r} \frac{\partial v}{\partial P_r} - \frac{\partial u}{\partial P_r} \frac{\partial v}{\partial Q_r})$$

$$= \sum_{r=1}^{m} (\frac{\partial u}{\partial Q_r} \frac{\partial v}{\partial P_r} - \frac{\partial u}{\partial P_r} \frac{\partial v}{\partial Q_r})$$

In other words, the Dirac bracket calculates the Poisson bracket based on only the unconstrained positions and momenta.

The extended Kalman filter in quantum mechanics for observables. Let

$$H = P^2/2 + U(Q)$$

The Belavkin filter in the vacuum coherent state is given by

$$d\pi_t(X) = \pi_t(\theta(X))dt + (\pi_t(XL^* + LX) - \pi_t(X)\pi_t(L + L^*))(dY_o(t) - \pi_t(L + L^*)dt)$$

where

$$\theta(X) = i[H, X] - (1/2)(LL^*X + XLL^* - 2LXL^*)$$

$$= i[H, X] - (1/2)(L[L^*, X] + [X, L]L^*)$$

We choose

$$L = g(Q)$$

where g is a real valued function. Then in the position representation

$$\rho(t, Q, Q') = \int W(t, (Q + Q')/2, P)exp(iP(Q - Q')/h)dP,$$

$$W(t, Q, P) = \int \rho(t, Q + q/2, Q - q/2)exp(iPq)dq$$

W is the Wigner distribution. We calculate

$$Tr(\rho X) = \int \rho(Q, Q')X(Q', Q)dQdQ'$$

$$= \int W((Q + Q')/2, P)exp(iP(Q - Q')/h)X(Q', Q)dPdQdQ'$$

$$= \int W(Q, P)exp(2iPq/h)X(Q + q, Q - q)dPdQdq$$

$$= \int W(Q, P)exp(2iPq/h)Y(Q, q)dPdQdq$$

(where $Y(Q, q) = X(Q+q, Q-q)$ or equivalently, $X(Q, Q') = Y((Q+Q')/2, (Q - Q')/2)$). Now, this can be expressed as (after putting in a normalizing factor $1/\pi h$

$$(\pi h)^{-1}Tr(\rho X) = \sum_{n \geq 0} \int W(Q, P)Y_n(Q)q^n exp(2iPq/h)dPdQdq/\pi h$$

$$= \int W(Q, P)Y_0(Q)exp(2iPq/h)dPdQdq/\pi h$$

$$+ \sum_{n \geq 1}(-h/2i)^n(\partial_P^n W(Q, P))Y_n(Q)exp(2iPq/h)dPdQdq/\pi h$$

$$= \int W_1(Q)Y_0(Q)dQ + \sum_{n \geq 1}(ih/2)^n \int (\partial_P^n W(Q, 0))Y_n(Q)dQ$$

Note that

$$Y_0(Q) = Y(Q, 0) = X(Q, Q)$$

is the diagonal slice of the observable X in the position space representation and

$$Y(Q, q) = X(Q + q, Q - q) = \sum_{n \geq 0} Y_n(Q)q^n$$

so that

$$W_1(Q) = \int W(Q,P)dP$$

is the marginal probability density of Q. Further, for an operator

$$M = \sum_{n \geq 0} g_n(Q)P^n$$

we have

$$MXf(Q) = \sum_n \int g_n(Q)(-ih\partial_Q)^n X(Q,Q')f(Q')dQ'$$

and hence,

$$Tr(\rho MX) = \sum_n \int \rho(Q,Q')g_n(Q')(-ih\partial'_Q)^n X(Q',Q)dQ'dQ$$

$$= \sum_n \int (ih\partial_{Q'})^n (W((Q+Q')/2,P)exp(iP(Q-Q')/h)g_n(Q'))X(Q',Q)dQ'dQ$$

$$= \sum_{n,k,r} (ih)^n 2^{-k} \binom{n}{k} \int (\partial_Q^k W(Q,P))(-iP/h)^{n-k} exp(2iPq/h)Y_r(Q)q^r dPdQdq$$

$$= \pi h \sum_{nkr} (ih)^n (ih/2)^r (-i/h)^{n-k} 2^{-k} \binom{n}{k} \int (\partial_P^r P^{n-k} \partial_Q^k W(Q,P))|_{P=0} Y_r(Q)dPdQdq$$

5.6   Harmonic oscillator with time varying electric field and Lindblad noise with Lindblad operators being linear in the creation and annihilation operators, transforms a Gaussian state into another after time t

Hamiltonian:

$$H(t) = \omega(t)a^*a + f(t)(a+a^*)$$

$$L = \alpha(t)a + \beta(t)a^*$$

Heisenberg equation of motion:

$$X'(t) = i[H(t), X(t)] + \theta_t(X(t)) = T_t(X(t))$$

where

$$\theta_t(X) = (-1/2)(LL^*X + XLL^* - 2LXL^*)$$
$$= (-1/2)(L[L^*, X] + [X, L]L^*)$$

and hence
$$T_t(X) = i[H(t), X] + \theta_t(X)$$

Let $\rho(t)$ be a Gaussian state. Its quantum Fourier transform is
$$\hat{\rho}(t, z) = Tr(\rho(t)W(z))$$

At time $t + dt$ the state is
$$\rho(t + dt) = \rho(t) + dt.T_t^*(\rho(t))$$

Its quantum Fourier transform is
$$\hat{\rho}(t + dt, z) = Tr(\rho(t + dt)W(z)) =$$
$$\hat{\rho}(t, z) + dt Tr(T_t^*(\rho(t))W(z))$$
$$= \hat{\rho}(t, z) + dt.Tr(\rho(t)T_t(W(z)))$$

Now,
$$T_t(W(z)) = i[H(t), W(z)] + \theta_t(W(z))$$

and
$$[H(t), W(z)] = [\omega(t)a^*a, W(z)] + f(t)[a + a^*, W(z)]$$

Now,
$$[a, W(z)] = zW(z), [a^*, W(z)] = \bar{z}W(z)$$

Then,
$$[a^*a, .W(z) = a^*[a, W(z)] + [a^*, W(z)]a =$$
$$za^*W(z) + \bar{z}W(z)a$$
$$= (za^* + \bar{z}a - |z|^2)W(z)$$

Thus,
$$[H(t), W(z)] = [\omega(t)(za^* + \bar{z}a - |z|^2) + f(t)(z + \bar{z})]W(z)$$

Further,
$$[L^*, W(z)] = [\bar{\alpha}(t)a^* + \bar{\beta}(t)a, W(z)] =$$
$$(\bar{\alpha}(t)\bar{z} + \bar{\beta}(t)z)W(z)$$
$$[W(z), L] = -(\alpha z + \beta \bar{z})W(z)$$

So
$$L[L^*, W(z)] + [W(z), L]L^* = (\alpha a + \beta a^*)(\bar{\alpha}\bar{z} + \bar{\beta}z)W(z) - (\alpha z + \beta \bar{z})W(z)(\bar{\alpha}a^* + \bar{\beta}a)$$
$$= (\alpha a + \beta a^*)(\bar{\alpha}\bar{z} + \bar{\beta}z)W(z) - (\alpha z + \beta \bar{z})([W(z), L^*] + L^*W(z))$$
$$= (\alpha a + \beta a^*)(\bar{\alpha}\bar{z} + \bar{\beta}z)W(z) + (\alpha z + \beta \bar{z})(\bar{\alpha}\bar{z} + \bar{\beta}z)W(z) - (\alpha z + \beta \bar{z})(\bar{\alpha}a^* + \bar{\beta}a)W(z)$$
$$= \lambda(\bar{z}a - za^*)W(z) - |\alpha z + \bar{z}|^2 W(z)$$

where
$$\lambda = |\alpha|^2 - |\beta|^2$$

5.7 Quantum neural network using a single harmonic oscillator perturbed by an electric field

Simulation studies.

The Hamiltonian is

$$H(t) = a^*a + f(t)(a+a^*) = b(t)^*b(t) - f(t)^2, b(t) = a + f(t), b(t)^* = a^* + f(t),$$

$$a = (q+ip)/\sqrt{2}, a^* = (q-ip)/\sqrt{2}$$

$$b(t)|0>_t = 0$$

implies

$$(q + \partial_q + f(t)\sqrt{2}) < q|\mathbf{0}>_t = 0$$

Then,

$$< q|\mathbf{0}>_t = C(t)exp(-q^2/2 - f(t)\sqrt{2}q) = \pi^{-1/4}.exp(-(q + f(t)\sqrt{2})^2/2)$$

Define

$$|n>_t = b(t)^{*n}|\mathbf{0}>_t /\sqrt{n!}$$

Then,

$$\partial_t|n>_t = (b(t)^{*n}/\sqrt{n!})(-f'(t)\sqrt{2}(q + f(t)\sqrt{2}))|0>_t + f'(t)\sqrt{n}|n-1>_t$$

Now,

$$q|0>_t = (a+a^*)|0>_t /\sqrt{2} = (b(t)+b(t)^*-2f(t))|0>_t /\sqrt{2} = (b(t)^*-2f(t))|0>_t /\sqrt{2}$$

Thus,

$$\partial_t|n>_t = -2f'(t)f(t)|n>_t - f'(t)(b(t)^{*n+1}/\sqrt{n!})|0>_t +2\sqrt{2}f(t)f'(t)|n>_t +\sqrt{n}f'(t)|n-1>_t$$

$$-2f'(t)f(t)|n>_t -\sqrt{n+1}f'(t)|n+1>_t +2\sqrt{2}f(t)f'(t)|n>_t +\sqrt{n}f'(t)|n-1>_t$$

$$= (2\sqrt{2} - 2)f(t)f'(t)|n>_t -\sqrt{n+1}f'(t)|n+1>_t +\sqrt{n}f'(t)|n-1>_t$$

$$= \alpha f(t)f'(t)|n>_t +\beta(n)f'(t)|n+1>_t +\gamma(n)f'(t)|n-1>_t$$

Schrodinger's equation gives for the wave function

$$i\partial_t|\psi(t)> = H(t)|\psi(t)>$$

Let

$$|\psi(t)> = \sum_n c(n,t)|n>_t$$

Then Schrodinger's equation translates in the coefficient domain to

$$i\partial_t c(n,t) + ic(n,t)\alpha f(t)f'(t) + \beta(n-1)if'(t)c(n-1,t) + \gamma(n+1)if'(t)c(n+1,t)$$

$$= (n - f(t)^2)c(n,t)$$

Let $p_0(t, q)$ be the pdf to be tracked. At time t, compute

$$c_0(n, t) = \int \sqrt{p_0(t, q)} < q | n >_t^* dq$$

We apply the LMS algorithm to track $c_0(n, t)$ using $c(n, t)$ by adapting the electric field $f(t)$. We write in discretized notation,

$$c(n, t+h) = c(n, t) + h(-i(n - f(t)^2) - \alpha f(t) f'(t)) c(n, t)$$

$$-\beta(n-1) f'(t) c(n-1, t) - \gamma(n+1) f'(t) c(n+1, t))$$

Writing

$$f'(t) = (f(t) - f(t-h))/h$$

and retaining only the dominant O(1) terms, ie, neglecting $O(h)$ terms on the rhs gives us

$$c(n, t+h) = c(n, t) - \alpha f(t)^2 c(n, t) - \beta(n-1) f(t) c(n-1, t) - \gamma(n+1) f(t) c(n+1, t)$$

Thus,

$$\partial c(n, t+h)/\partial f(t) = -2\alpha f(t) c(n, t) - \beta(n-1) c(n-1, t) - \gamma(n+1) c(n+1, t)$$

and hence the new weight $f(t+h)$ is given by the LMS algorithm:

$$f(t+h) = f(t) - \mu(\partial/\partial f(t)) \sum_n |c_0(n, t+h) - c(n, t+h)|^2$$

$$= f(t) + 2\mu \sum_n Re[(\bar{c}_0(n, t+h) - \bar{c}(n, t+h)) \partial c(n, t+h)/\partial f(t)]$$

$$= f(t) + 2\mu \sum_n Re[(\bar{c}_0(n, t+h) - \bar{c}(n, t+h))(-2\alpha f(t) c(n, t) - \beta(n-1) c(n-1, t) -$$

$$\gamma(n+1) c(n+1, t))]$$

Suppose however we do not neglect the non-dominant terms. Then, we would use the LMS algorithm with delay terms as follows:

$$f(t+h) = f(t) - \mu_1(\partial/\partial f(t)) \sum_n |c_0(n, t+h) - c(n, t+h)|^2$$

$$-\mu_2(\partial/\partial f(t-h)) \sum_n |c_0(n, t+h) - c(n, t+h)|^2$$

$$= f(t) + 2\mu_1 Re \sum_n [(\bar{c}_0(n, t+h) - \bar{c}(n, t+h))(\partial c(n, t+h)/\partial f(t))]$$

$$+ 2\mu_2 Re \sum_n [(\bar{c}_0(n, t+h) - \bar{c}(n, t+h))(\partial c(n, t+h)/\partial f(t-h))]$$

Now,

$$\partial c(n, t+h)/\partial f(t) =$$

$$(\partial/\partial f(t))(c(n,t) + h(-i(n - f(t)^2) - \alpha f(t)(f(t) - f(t - h))/h)c(n,t)-$$
$$\beta(n - 1)(f(t) - f(t - h))c(n - 1, t)/h - \gamma(n + 1)(f(t) - f(t - h))c(n + 1, t)/h))$$
$$= (2ihf(t) - \alpha((2f(t) - f(t - h)))c(n,t) - \beta(n - 1)c(n - 1, h) - \gamma(n+1)c(n+1, t)$$
$$\partial c(n, t + h)/\partial f(t - h) =$$
$$\alpha f(t)c(n, t) + \beta(n - 1)c(n - 1, t) + \gamma(n + 1)c(n + 1, t)$$

The LMS algorithm can be developed along these lines.

Chapter 6

Statistical Signal Processing

6.1   Statistical Signal Processing: Long Test

Answer all questions. Each question carries ten marks.

[1] Let $X(n), n \geq 0$ be a Markov chain in discrete time with finite state space $E = \{1, ..., N\}$ and transition probability

$$Pr(X(n+1) = j | X(n) = i) = \pi(i, j | \theta), i, j = 1, 2, ..., N$$

Let $X(0)$ have the probability distribution $p_0(i), i = 1, 2, ..., N$. Write down the expression for the joint probability distribution of $X(1), ..., X(K)$ given θ and assuming that the maximum likelihood estimator $\hat{\theta} = \theta_0 + \delta\theta$ is a small perturbation of a known value θ_0, evaluate $\delta\theta$ by expanding the likelihood function upto quadratic orders in $\delta\theta$. Also write down an expression for the Cramer-Rao lower bound on the variance of an unbiased estimate $\delta\theta$ of the parameter perturbation from θ_0.

[2] Consider Schrodinger's equation for the wave function $\psi(t, \mathbf{r})$ of an electron when a random electromagnetic field dependent upon a parameter vector θ is incident upon the electron. The random electromagnetic field is specified by a magnetic vector potential $\mathbf{A}(t, \mathbf{r}|\theta)$ and an electric scalar potential $\Phi(t, \mathbf{r}|\theta)$. Assume that these have the expansions

$$\mathbf{A}(t, \mathbf{r}|\theta) = \sum_{k=1}^{p} \theta_k \mathbf{A}_k(t, \mathbf{r}),$$

$$\Phi(t, \mathbf{r}|\theta) = \sum_{k=1}^{p} \theta_k \Phi_k(t, \mathbf{r})$$

where $\mathbf{A}_k, \Phi_k$ are zero mean Gaussian random fields with known autocorrelation functions. The Hamiltonian of the electromagnetically perturbed electron is given by

$$H(t) = (-h^2/2m)(\nabla + ir\mathbf{A}(t, \mathbf{r}|\theta)/h)^2 - Ze^2/|\mathbf{r}| - e\Phi(t, \mathbf{r}|\theta)$$

Express this Hamiltonian as

$$H(t) = H_0 + eV_1(t|\theta) + e^2 V_2(t|\theta)$$

where

$$H_0 = (-h^2/2m)\nabla^2 - Ze^2/|\mathbf{r}|$$

is the Hamiltonian of the unperturbed electron bound to its nucleus. Using second order perturbation theory, approximately calculate the mixed state of the electron

$$\rho(t, \mathbf{r}, \mathbf{r}') = \mathbb{E}(\psi(t, \mathbf{r})\bar{\psi}(t, \mathbf{r}'))$$

at time t upto second order in θ in terms of the autocorrelations of the fields $\mathbf{A}_k, \Phi_k, k = 1, 2, ..., p$. Hence derive the maximum likelihood estimator of θ based on making measurements at discrete times $t_1 < t_2 < .. < t_N$ using a projection valued measure $\{\mathbf{M}_a : a = 1, 2, ..., r\}$ taking into account the collapse postulate following each measurement.

[3] Show that if $X(n), n = 0, 1, ...$ is a Markov process in discrete time with finite state space $\{1, 2, ..., A\}$, then the large deviation rate function for the empirical distribution of the process based on time averages is given by

$$I(\mathbf{q}) = sup_{u>0} \sum_{k=1}^{A} q(k) log(u(k)/\pi u(k))$$

where

$$\pi u(k) = \sum_{j=1}^{A} \pi(k, j) u(j)$$

with $\pi(k, j)$ denoting the one step transition probability distribution of the process. Evaluate using this and a limiting scheme in which a continuous time Markov process is approximated by a discrete time Markov process, the rate functions for the empirical density for a diffusion process $X(t)$ defined by the equation

$$dX(t) = \mu.dt + \sigma.dB(t)$$

where $B(t)$ is standard Brownian motion and μ, σ are non-random parameters. If μ and σ are functions of an unknown parameter θ, then explain how to estimate θ so that the empirical distribution of the $X(t)$ process matches as closely as possible a given probability distribution in the mean square sense.

[4] Given a vector valued iid $N(0, \mathbf{R})$ process $\mathbf{X}(n), n = 0, 1, ...$ in discrete time and a scalar process

$$d(n) = \mathbf{h}^T \mathbf{X}(n) + v(n), n = 0, 1, ...$$

where $v(n)$ is iid $N(0, \sigma_v^2)$ independent of the $X(n)$ process, explain how you will estimate $\mathbf{h}$ using the LMS algorithm. Also perform a convergence analysis of the weight vector by determining its asymptotic mean and covariance in terms of $\mathbf{h}, \mathbf{R}, \sigma_v^2$. How will you generalize this to the case when $\mathbf{h}$ is a matrix, and $v(n)$ is a vector so that $d(n)$ also becomes a vector ?

[5] [a] Show that if $0 \leq \mu \leq 1$, then the map

$$(X, Y) \to X^{1-\mu} \otimes Y^\mu$$

where X and Y vary over the space of $n \times n$ complex positive definite matrices is a concave function and using this deduce that if ρ, σ are any two quantum states in the Hilbert space $\mathbb{C}^n$ and K is any quantum pinching, ie, of the form

$$K(\rho) = \sum_{i=1}^{d} P_i \rho P_i$$

where $\{P_i\}$ is an orthogonal resolution of the identity, then

$$D(K(\rho)|K(\sigma)) \leq D(\rho|\sigma)$$

where

$$D(\rho|\sigma) = Tr(\rho.(log(\rho) - log(\sigma)))$$

is the quantum relative entropy.

hint: Use the fact that any quantum pinching can be represented as a finite composition of pinchings of the form

$$K(\rho) = (\rho + U\rho U^*)/2$$

where U is a unitary operator.

[b] Let f be a matrix convex function, ie, for any two positive matrices A, B of the same size, we have

$$f(tA + (1-t)B) \leq tf(A) + (1-t)f(B), 0 \leq t \leq 1$$

Then, prove that if $\mathbf{Z}_1, ..., \mathbf{Z}_d$ are matrices of the same size such that

$$\sum_{k=1}^{d} \mathbf{Z}_k^* \mathbf{Z}_k \leq \mathbf{I}$$

we have

$$f(\sum_{k=1}^{d} Z_k^* A Z_k) \leq \sum_{k=1}^{d} Z_k^* f(A) Z_k$$

for any positive matrix A. Using this and the operator convex function $x \to -x^{1-s}$ for $x > 0$ and $0 \leq s \leq 1$, deduce that the quantum relative Renyi entropy between two states defined by

$$D_s(\rho|\sigma) = -s^{-1}log(Tr(\rho^{1-s}\sigma^s))$$

satisfies the monotone property, ie,

$$D_s(K(\rho)|K(\sigma)) \leq D_s(\rho|\sigma), 0 \leq s \leq 1$$

where K is any quantum operation, ie, a TPCP map.

hint: By defining the operator $\Delta_{\rho,\sigma} = R_\rho L_\sigma^{-1}$ so that $\Delta_{\rho,\sigma}(X) = \sigma^{-1}X\rho$, choose an appropriate inner product on the space of matrices so that $\Delta_{\rho,\sigma}$ becomes a positive operator. Then define $K_{\sigma,r}(X) = K(\sigma)^{-1}K(\sigma X)$ where K is partial trace. Show that the dual of the linear operator $K_{\sigma,r}$, is given by $K_{\sigma,r}^*(Y) = Y \otimes I$ provided that we select appropriate inner products on the domain and range space of $K_{\sigma,r}$. Then, deduce using operator convexity of $f(x) = -x^{1-s}$ that $0 \leq s \leq 1$, we have that

$$Tr(K(\rho)^{1-s}K(\sigma^s)) \geq Tr(\rho^{1-s}\sigma^s) - - - (1)$$

For doing this problem, you must make use of the result just stated above in the form

$$f(K_{\sigma,r}\rho K_{\sigma,r}^*) \leq K_{\sigma,r}f(\rho)K_{\sigma,r}^*$$

Now use the fact that if K is an arbitary TPCP map, then it has the Stinespring representation

$$K(\rho) = Tr_2(U(\rho \otimes \rho_0)U^*)$$

where ρ_0 is a state and U is a unitary operator. Using the inequality already established for the partial trace, deduce the same inequality for general TPCP K.

6.2   Quantum EKF

Let $X(Q, Q')$ be the position space representation of an observable X and let $\rho(Q, Q')$ be the position space representation of a density matrix ρ. The average of X in the state ρ,

$$Tr(\rho.X) = \int \rho(Q, Q')X(Q', Q)dQdQ'$$

Let $W(Q, P)$ denote the Wigner distribution of ρ:

$$W(Q, P) = C \int \rho(Q + q/2, Q - q/2)exp(-iPq/h)dq$$

where C is a normalization constant that ensures

$$\int W(Q, P)dQdP = 1$$

Thus, since

$$\int exp(-iPq/h)dP = 2\pi h\delta(q)$$

it follows that

$$2\pi hC \int \rho(Q,Q)dQ = 1$$

or

$$C = 1/2\pi h$$

Then,

$$\rho(Q,Q') = \int W((Q+Q')/2, P)exp(iP(Q-Q')/h)dP$$

We find that

$$\rho(Q,Q) = \int W(Q,P)dP,$$

$$\int W(Q,P)dQ = C \int \rho(Q+q/2, Q-q/2)exp(-iPq/h)dqdQ$$

$$= (2\pi h)^{-1} \int \rho(Q,Q')exp(-iP(Q-Q')/h)dQdQ'$$

Thus, $\int W(Q,P)dP$ is the probability density of Q while $\int W(Q,P)dQ$ is the probability density of P in the state ρ. The average value of X is

$$Tr(\rho X) = \int W((Q+Q')/2, P)exp(iP(Q-Q')/h)X(Q',Q)dPdQdQ'$$

$$= \int W(Q,P)X_1(Q,P)dQdP$$

where X_1 is the Wigner transform of X defined by

$$X_1(Q,P) = \int X(Q-q/2, Q+q/2)exp(iPq/h)dq$$

$$= \int X(Q+q/2, Q-q/2)exp(-iPq/h)dq$$

Now we are in a position to derive the EKF for observables in the quantum theory using the Belavkin filter. The Belavkin quantum filter equation for $L = aQ + bP$ piin a vaccum coherent state with the particle moving in a potential $U(Q)$ has the form described earlier but we shall rederive it here:

$$\theta(X) = i[H,X] - (1/2)(LL^*X + XLL^* - 2LXL^*)$$

$$d\pi_t(X) = \pi_t(\theta(X))dt + (\pi_t(XL^*+LX) - \pi_t(X)\pi_t(L+L^*))(dY(t) - \pi_t(L+L^*)dt)$$

Now,

$$\pi_t(\theta(X)) = \int W(t,Q,P)\theta(X)_1(Q,P)dQdP$$

$$\theta(X)_1(Q,P) = [H,X]_1(Q,P) - (1/2)(L[L^*,X] + [X,L]L^*)_1(Q,P)$$

$$LL^*X = (aQ+bP)(\bar{a}Q+\bar{b}P)X = (|a|^2Q^2 + a\bar{b}QP + \bar{a}bPQ + |b|^2P^2)X$$

For convenience, let $b = 0$, ie, $L = aQ$. Then

$$LL^*X = |a|^2Q^2X, XLL^* = |a|^2XQ^2, LXL^* = |a|^2QXQ$$

and hence,

$$(LL^*X)_1(Q,P) = |a|^2 \int (Q+q/2)^2 X(Q+q/2, Q-q/2)exp(iPq/h)dq$$

$$= |a|^2[Q^2 X_1(Q,P) + Q(h/i)\partial_P X_1(Q,P) + (1/4)(h/i)^2\partial_P^2 X_1(Q,P)]$$

$$(XLL^*)_1(Q,P) = |a|^2 \int X(Q+q/2, Q-q/2)(Q-q/2)^2 exp(iPq/h)dq$$

$$|a|^2[Q^2 X_1(Q,P) - Q(h/i)\partial_P X_1(Q,P) + (1/4)(h/i)^2\partial_P^2 X_1(Q,P)]$$

$$(LXL^*)_1(Q,P) = |a|^2[\int (Q+q/2)(Q-q/2)X(Q+q/2, Q-q/2)exp(iPq/h)dq]$$

$$= |a|^2[Q^2 X_1(Q,P) - (1/4)(h/i)^2\partial_P^2 X_1(Q,P)]$$

Thus,

$$(LL^*X + XLL^* - 2LXL^*)_1(Q,P) =$$
$$|a|^2(h/i)^2\partial_P^2 X_1(Q,P) = -|a|^2 h^2 \partial_P^2 X_1(Q,P)$$
$$[H,X]_1(Q,P) = [P^2/2, X]_1(Q,P) + [U(Q), X]_1(Q,P)$$

Now,

$$(P^2 X)_1(Q,P) = \int (-h^2)\partial_1^2 X(Q+q/2, Q-q/2)exp(iPq/h)dq$$

Now,

$$(PX)_1(Q,P) = -ih\int (\partial_Q/2 + \partial_q)X(Q+q/2, Q-q/2)exp(iPq/h)dq$$

$$= (-ih)[\partial_Q X_1(Q,P) - (iP/h)X_1(Q,P)]$$

Applying this operator twice gives

$$(P^2 X)_1(Q,P) = -h^2(\partial_Q - iP/h)^2 X_1(Q,P)$$

6.3 Lie brackets in quantum mechanics in terms of the Wigner transform of observables

Identifying the quantum corrections in powers of Planck's constant that the Lie bracket gives to the classical Poisson bracket. Let X,Y be two observables with position space representations $X(Q,Q'), Y(Q,Q')$ respectively. Their respective Wigner transforms are

$$X_1(Q,P) = \int X(Q+q/2, Q-q/2)exp(-iPq/h)dq,$$

$$Y_1(Q, P) = \int Y(Q + q/2, Q - q/2) exp(-iPq/h) dq$$

We have

$$(XY)_1(Q, P) = \int (XY)(Q + q/2, Q - q/2) exp(-iPq/h) dq$$

$$= \int X(Q + q/2, Q') Y(Q', Q - q/2) exp(-iPq/h) dq dQ'$$

$$= \int X_1((Q+Q')/2+q/4, P_1) exp(iP_1(Q-Q'+q/2)/h) Y_1$$

$$((Q+Q')/2-q/4, P_2) exp(iP_2(Q'-Q+q/2)/h) exp(-iPq/h) dP_1 dP_2 dq dQ'$$

We write

$$X_1(Q + q, P) = \sum_{n \geq 0} X_n(Q, P) q^n, Y_1(Q + q, P) = \sum_{n \geq 0} Y_n(Q, P) q^n$$

Then,

$$(XY)_1(Q, P) = \sum_{n,m \geq 0} \int \tilde{X}_n((Q+Q')/2, P_1) \tilde{Y}_m((Q+Q')/2, P_2)(-1)^m (q/4)^{n+m}$$

$$.exp(i(P_1 - P_2)(Q - Q')/h) exp(i((P_1 + P_2)/2 - P) q/h) dP_1 dP_2 dq dQ'$$

$$= \sum_{n,m} (-1)^m \int \tilde{X}_n((Q+Q')/2, P_1) \tilde{Y}_m((Q+Q')/2, P_2)$$

$$exp(i(P_1-P_2)(Q-Q')/h)(h/4i)^{n+m} \delta^{(n+m)}((P_1+P_2)/2-P) dP_1 dP_2 dQ'$$

$$= \sum_{n,m} (-1)^m (ih/4)^{n+m} \partial_P^{n+m} \int \tilde{X}_n(Q+q/2, P_1) \tilde{Y}_m(Q+q/2, P_2)$$

$$exp(-i(P_1-P_2)q/h) \delta(P-(P_1+P_2)/2) dP_1 dP_2 dq$$

$$= \sum_{n,m} (-1)^m (ih/4)^{n+m} \partial_P^{n+m} \int \tilde{X}_n(Q+q/2, P_0+p/2) \tilde{Y}_m(Q+q/2, P_0-p/2) exp(-ipq/h) \delta(P-P_0) dP_0 dp dq$$

$$= \sum_{n,m} (-1)^m (ih/4)^{n+m} \partial_P^{n+m} \int \tilde{X}_n(Q+q/2, P+p/2) \tilde{Y}_m(Q+q/2, P-p/2) exp(-ipq/h) dp dq$$

$$= h \sum_{n,m} (-1)^m (ih/4)^{n+m} \partial_P^{n+m} \int \tilde{X}_n(Q+hq/2, P+p/2) \tilde{Y}_m(Q+hq/2, P-p/2) exp(-ipq) dp dq$$

Noting that
The coefficient of h^2 in this expansion is

$$(i/4)\partial_P \int (\tilde{X}_1(Q, P+p/2)\tilde{Y}_0(Q, P-p/2) - \tilde{X}_0(Q, P+p/2)\tilde{Y}_1$$

$$(Q, P-p/2)) exp(-ipq) dp dq$$

$$+ (1/2) \int (\tilde{X}_{0,1}(Q, P+p/2)\tilde{Y}_0(Q, P-p/2) + \tilde{X}_0(Q, P+p/2)\tilde{Y}_{0,1}$$

$$(Q, P-p/2)) q.exp(-ipq) dp dq$$

$$= (i\pi/2)\partial_P(\tilde{X}_1(Q, P)\tilde{Y}_0(Q, P) - \tilde{X}_0(Q, P)\tilde{Y}_1(Q, P))$$

$$+(i\pi/2)(\tilde{X}_{0,12}(Q,P)\tilde{Y}_0(Q,P) - \tilde{X}_{0,1}(Q,P)\tilde{Y}_{0,2}(Q,P)$$
$$+\tilde{X}_{0,2}(Q,P)\tilde{Y}_{0,1}(Q,P) - \tilde{X}_0(Q,P)\tilde{Y}_{0,12}(Q,P))$$
$$= (i\pi/2)\partial_P(X_{1,1}(Q,P)Y_1(Q,P) - X_1(Q,P)Y_{1,1}(Q,P))$$
$$-(i\pi/2)(X_{1,12}(Q,P)Y_1(Q,P) - X_{1,1}(Q,P)Y_{1,2}(Q,P) + X_{1,2}(Q,P)Y_{1,1}(Q,P)$$
$$-X_1(Q,P)Y_{1,12}(Q,P))$$

6.4   Simulation of a class of Markov processes in continuous and discrete time with applications to solving partial differential equations

[1] Simulation of Brownian motion.
   [2] Simulation of Brownian motion with drift.
   [3] Simulation of the Poisson process.
   [4] Simulation of the Birth-death process.
   [5] Simulation of a Markov chain in continuous time with finite state space and prescribed infinitesimal generator.
   [6] Simulation of diffusion processes driven by Brownian motion.
   [7] Simulation of independent increment processes with prescribed characteristic function based on compound Poisson processes.
   Chapter: Classical and Quantum Information Theory

6.5   Gravitational radiation

$$G^{\mu\nu} = R^{\mu\nu} - (1/2)Rg^{\mu\nu}$$

is the Einstein tensor. Its covariant divergence vanishes:

$$G^{\mu\nu}_{:\nu} = 0$$

We can express $G^{\mu\nu}$ as the sum of a linear part and a nonlinear part in the metric perturbations $h_{\mu\nu}$ and its partial derivatives. Writing

$$g_{\mu\nu} = \eta_{\mu\nu} + h_{\mu\nu}$$

where $\eta_{\mu\nu}$ is the Minkowski tensor, we have

$$g^{\mu\nu} = \eta_{\mu\nu} - h^{\mu\nu}, h^{\mu\nu} = \eta_{\mu\alpha}\eta_{\nu\beta}h_{\alpha\beta}$$

Upto second order in the metric peturbations, we have

$$R_{\mu\nu} = (\eta_{\alpha\beta} - h^{\alpha\beta})\Gamma_{\beta\mu\alpha}),_\nu - ((\eta_{\alpha\beta} - h^{\alpha\beta})\Gamma_{\beta\mu\nu}),_\alpha$$

$$-\eta_{\alpha\rho}\eta_{\beta\sigma}\Gamma_{\rho\mu\nu}\Gamma_{\sigma\alpha\beta} + \eta_{\alpha\rho}\eta_{\beta\sigma}\Gamma_{\rho\mu\beta}\Gamma_{\sigma\nu\alpha}$$

where

$$\Gamma_{\beta\mu\alpha} = (1/2)(h_{\beta\mu,\alpha} + h_{\beta\alpha,\mu} - h_{\alpha\mu,\beta})$$

Upto linear orders in the metric perturbations, we have

$$R_{\mu\nu} = (1/2)\eta_{\alpha\beta}(h_{\beta\mu,\alpha} + h_{\beta\alpha,\mu} - h_{\mu\alpha,\beta})_{,\nu}$$

$$-(1/2)\eta_{\alpha\beta}(h_{\beta\mu,\nu} + h_{\beta\nu,\mu} - h_{\mu\nu,\beta})_{,\alpha})$$

$$= (1/2)(h_{,\mu\nu} - h^{\alpha}_{\mu,\nu\alpha} - h^{\alpha}_{\nu,\mu\alpha} + \Box h_{\mu\nu})$$

On imposing the coordinate condition

$$g^{\mu\nu}\Gamma^{\alpha}_{\mu\nu} = 0$$

and linearizing it, we get

$$(1/2)h_{,\alpha} - h^{\nu}_{\alpha,\nu} = 0$$

and hence under this condition, we get the linear component of $R_{\mu\nu}$ as

$$R^{(1)}_{\mu\nu} = (1/2)\Box h_{\mu\nu}$$

The quadratic component of $R_{\mu\nu}$ evaluates to

$$R^{(2)}_{\mu\nu} = -(h^{\alpha\beta}\Gamma_{\beta\mu\alpha})_{,\nu} + (h^{\alpha\beta}\Gamma_{\beta\mu\nu})_{,\alpha}$$

$$-\eta_{\alpha\rho}\eta_{\beta\sigma}\Gamma_{\rho\mu\nu}\Gamma_{\sigma\alpha\beta} + \eta_{\alpha\rho}\eta_{\beta\sigma}\Gamma_{\rho\mu\beta}\Gamma_{\sigma\nu\alpha}$$

Time averaged power radiated out by a matter source in the form of gravitational radiation. The Einstein field equations can be expressed as

$$R^{\mu\nu} = K.S^{\mu\nu}, S^{\mu\nu} = T^{\mu\nu} - (1/2)Tg^{\mu\nu}$$

where $T^{\mu\nu}$ is the energy-momentum tensor of all the matter and radiation fields. Upto linear orders in the metric perturbation, it can be expressed as

$$R^{\mu\nu(1)} = KS^{\mu\nu}$$

or equivalently after adopting the harmonic coordinate condition,

$$\Box h^{\mu\nu} = 2KS^{\mu\nu}$$

For sinusoidally varying matter fields, we have in the frequency domain,

$$h_{\mu\nu}(\omega, r) = (-K/2\pi) \int S^{\mu\nu}(\omega, r') exp(-jw|r - r'|)d^3r'/|r - r'|$$

which in the far field approximation, gives

$$h_{\mu\nu}(\omega, r) = (-K/2\pi)(exp(-j\omega r)/r) \int S^{\mu\nu}(\omega, r') exp(j\omega \hat{r}.r') d^3 r'$$

Let us now evaluate the time averaged power radiated out by the matter source at a given frequency. First, we observe that

$$G^{\mu\nu(2)} = R^{\mu\nu(2)} - (1/2)R^{(2)}\eta_{\mu\nu} + (1/2)R^{(1)}h^{\mu\nu}$$

where

$$R^{2)} = (g^{\mu\nu}R_{\mu\nu})^{(2)} = \eta_{\mu\nu}R_{\mu\nu}^{(2)} + h^{\mu\nu}R_{\mu\nu}^{(1)}$$

Next we observe that since

$$G_{,\nu}^{\mu\nu(1)} = 0$$

it follows that

$$(KT^{\mu\nu} - G^{\mu\nu(2)})_{,\nu} = 0$$

which means that the pseudo-tensor

$$Q^{\mu\nu} = KT^{\mu\nu} - G^{\mu\nu(2)}$$

is a conserved quantity. Specifically, the equation $Q_{,\nu}^{\mu\nu} = 0$ implies that

$$\frac{d}{dt} \int_V Q^{\mu 0} d^3 x = - \int_S Q^{\mu r} n_r dS$$

where V is any volume of three dimensional space bounded by a closed surface S whose unit outward normal is $n_r, r = 1, 2, 3$. This means that the rate at which the spatial integral of the total density Q^{00} of the energy in matter plus gravity over V increases with time equals the rate at which this total energy flows into V from its boundary. And likewise, the rate at which the spatial integral of the total density Q^{r0} of the momentum in matter plus gravity over V increases with time equals the rate at which this total momentum flows into V from its boundary. It follows from this that $-G^{\mu\nu(2)}$ must be interpreted as the energy-momentum pseudotensor of the gravitational field. The total rate at which the matter source within a region V of space bounded by a closed surface S emits energy in the form of gravitational radiation is therefore given by

$$- \int_S G^{0r(2)} n_r dS = - \int_V G_{,r}^{0r(2)} d^3 x$$

Gravitational radiation by a matter source in the far field zone.

$$R_{\mu\nu} = R_{\mu\nu}^{(1)} + R_{\mu\nu}^{(2)}$$

$$R^{\mu\nu} = R^{\mu\nu(1)} + R^{\mu\nu(2)}$$

$$R^{\mu\nu(1)} = (g^{\mu\alpha}g^{\nu\beta}R_{\alpha\beta})^{(1)}$$

$$= \eta_{\mu\alpha}\eta_{\nu\beta}R^{(1)}_{\alpha\beta}$$

$$R^{\mu\nu(2)} = (\eta_{\mu\alpha} - h^{\mu\alpha})(\eta_{\nu\beta} - h^{\nu\beta})(R^{(1)}_{\alpha\beta} + R^{(2)}_{\alpha\beta})$$

$$= \eta_{\mu\alpha}\eta_{\nu\beta}R^{2)}_{\alpha\beta}$$

$$-(\eta_{\mu\alpha}h^{\nu\beta} + \eta_{\nu\beta}h^{\mu\alpha})R^{(1)}_{\alpha\beta}$$

$$G^{\mu\nu(1)} = R^{\mu\nu(1)} - (1/2)R^{(1)}\eta_{\mu\nu}$$

$$R^{(1)} = \eta_{\mu\nu}R^{(1)}_{\mu\nu}$$

$$G^{\mu\nu(2)} = R^{\mu\nu(2)} - (1/2)R^{(2)}\eta_{\mu\nu} + (1/2)R^{1)}h^{\mu\nu}$$

$$R^{(2)} = (g^{\mu\nu}R_{\mu\nu})^{(2)} =$$

$$\eta_{\mu\nu}R^{(2)}_{\mu\nu} - h^{\mu\nu}R^{(1)}_{\mu\nu}$$

Thus,

$$G^{\mu\nu(2)} =$$

$$\eta_{\mu\alpha}\eta_{\nu\beta}R^{(2)}_{\alpha\beta}$$

$$-(\eta_{\mu\alpha}h^{\nu\beta} + \eta_{\nu\beta}h^{\mu\alpha})R^{(1)}_{\alpha\beta}$$

$$-(1/2)\eta_{\mu\nu}(\eta_{\alpha\beta}R^{(2)}_{\alpha\beta} - h^{\alpha\beta}R^{(1)}_{\alpha\beta})$$

$$+(1/2)h^{\mu\nu}\eta_{\alpha\beta}R^{(1)}_{\alpha\beta}$$

$$= (\eta_{\mu\alpha}\eta_{\nu\beta} - (1/2)\eta_{\mu\nu}\eta_{\alpha\beta})R^{(2)}_{\alpha\beta}$$

$$-(\eta_{\mu\alpha}h^{\nu\beta} + \eta_{\nu\beta}h^{\mu\alpha} - (1/2)(\eta_{\mu\nu}h^{\alpha\beta} + \eta_{\alpha\beta}h^{\mu\nu}))R^{(1)}_{\alpha\beta}$$

Flux of gravitational radiation:

$$P^r = -G^{0r(2)}, r = 1, 2, 3$$

It is clear that by excluding total differential, we can write

$$P^r = C(r\mu\nu\alpha\rho\sigma\beta)h_{\mu\nu,\alpha}h_{\rho\sigma,\beta}$$

where $C(.)$ are constants. Further, in the far field zone,

$$h_{\mu\nu}(\omega, r) = K.(exp(-i\omega r)/r)\int T_{\mu\nu}(\omega, r')exp(ik\hat{r}.r')d^3r'$$

$$= (exp(-i\omega r)/r)H_{\mu\nu}(\omega, \hat{r})$$

If we differentiate $H_{\mu\nu}(\omega, \hat{r})$ w.r.t x, y, z the result will be $O(1/r^2)$ terms in the $h_{\mu\nu,s}$ where $s = 1, 2, 3$. These terms cannot contribute to gravitational radiation in the far field zone since they contribute $O(1/r^3)$ terms to P^r. However, if we differentiate $exp(-i\omega r)$ w.r.t x, y, z, it will result in $O(1/r)$ terms in $h_{\mu\nu,s}$ which can contribute to gravitational radiation. Thus, the form of P^r in the far field zone in the frequency domain is

$$P^r(\omega, r) = \omega^2 Re(D(r, \mu\nu\alpha\beta) H_{\mu\nu}(\omega, \hat{r}) H_{\alpha\beta}(\omega, \hat{r})^* / r^2)$$

where the $D(.)'s$ are complex constants. The flux of gravitational radiation energy in the far field zone as a function of the direction is therefore

$$r^2 P^r(\omega, \hat{r}) =$$

$$\omega^2 Re(D(r, \mu\nu\alpha\beta) H_{\mu\nu}(\omega, \hat{r}) H_{\alpha\beta}(\omega, \hat{r})^*)$$

The gravitational energy radiated per unit time by a rotating pulsar. Assume that the pulsar centre is at the origin and that two masses each of mass m are rotating around this centre with a uniform angular velocity ω, both masses being at a distance d from the centre. To calculate the gravitational wave produced by this rotating pulsar, we must first evaluate the energy-momentum tensor corresponding to this pulsar. It is given by

$$T^{\mu\nu}(x) = m.(dx_1^\mu/d\tau_1)(dx_1^\nu/dt)\delta^3(x - x_1) + m.(dx_2^\mu/d\tau_1)(dx_2^\nu/dt)\delta^3(x - x_2)$$

In the cartesian system,

$$((x_1^r(t))) = d(cos(\omega t), sin(\omega t), 0),$$

$$((x_2^r(t))) = -d(cos(\omega t), sin(\omega t), 0),$$

$$dt/d\tau_1 = \gamma_1 = (1 - v_1^2)^{-1/2} = (1 - d^2\omega^2), = dt/d\tau_2 = \gamma_2 = \gamma$$

Note that our units are chosen so that $c = 1$. Then,

$$T^{00} = m\gamma_1 \delta^3(x - x_2) + m\gamma_2 \delta^3(x - x^2) =$$

$$= m\gamma(\delta^3(x - x_1) + \delta^3(x - x_2)),$$

$$T^{01} = T^{10} = m\gamma v_1^1 \delta^3(x - x_1) + m\gamma v_2^1 \delta^3(x - x_2)$$

$$= -m\gamma\omega d.sin(\omega t)(\delta^3(x - x_1) - \delta^3(x - x_2))$$

$$T^{02} = T^{20} = m\gamma v_1^2 \delta^3(x - x_1) + m\gamma v_2^2 \delta^3(x - x_2)$$

$$= m\gamma\omega d.cos(\omega t)(\delta^3(x - x_1) - \delta^3(x - x_2))$$

$$T^{03} = T^{30} = 0$$

$$T^{11} = m\gamma(v_1^1)^2 \delta^3(x - x_1) + m\gamma(v_2^1)^2 \delta^3(x - x_2)$$

$$T^{rs} = m\gamma v_1^r v_1^s \delta^3(x - x_1) + m\gamma v_2^r v_2^s \delta^3(x - x_2)$$
$$= m\gamma v^r v^s (\delta^3(x - x_1) + \delta^3(x - x_2)), r, s = 1, 2, 3$$

where

$$v^1 = -dw.sin(\omega t), v^2 = dw.cos(\omega t), v^3 = 0$$

We can abbreviate these expressions to

$$T^{\mu\nu}(x) = f^{\mu\nu}(t)\delta^3(x - x_1(t)) + k^{\mu\nu}(t)\delta^3(x - x_2(t))$$

Now we compute the metric perturbations generated by this energy-momentum tensor: First observe the simple fact that if we assume that only $f^{\mu\nu}(t)$ varied with time, ie, the masses of the two particles varied with time but their positions $x_1(t), x_2(t)$ did not vary with time, then we would derive the simple formulas

$$h_{\mu\nu}(\omega, r) = K \int T_{\mu\nu}(\omega, r') exp(-i\omega|r - r'|)d^3r'/|r - r'|$$

$$= K \int f^{\mu\nu}(\omega)\delta^3(r' - r_1) exp(-i\omega|r - r'|)d^3r'/|r - r'|$$

$$+K \int k^{\mu\nu}(\omega)\delta^3(r' - r_1) exp(-i\omega|r - r'|)d^3r'/|r - r'|$$

$$= K f^{\mu\nu}(\omega) exp(-i\omega|r - r_1|)/|r - r_1|$$

$$+K k^{\mu\nu}(\omega) exp(-i\omega|r - r_1|)/|r - r_1|$$

More generally, if we had N point masses moving rapidly in the vicinity of fixed points $r_1, ..., r_N$ in space respectively with three velocities $v_1^r(t), ..., v_N^r(t)$ and rest masses $m_1, ..., m_N$, then writing $v_k^0(t) = 1$ and $\gamma_k(t) = (1 - v_k(t)^2)^{-1/2}$ where $v_k(t)^2 = \sum_{r=1}^{3}(v_k^r(t))^2$, we would have the result that the approximate energy-momentum tensor of this system of particles is given by

$$T^{\mu\nu}(x) = T^{\mu\nu}(t, r) = \sum_{k=1}^{N} m_k \delta^3(r - r_k)\gamma_k(t)v_k^\mu(t)v_k^\nu(t)$$

$$\sum_{k=1}^{N} f_k^{\mu\nu}(t)\delta^3(r - r_k)$$

where

$$f_k^{\mu\nu}(t) = m_k\gamma_k(t)v_k^\mu(t)v_k^\nu(t), k = 1, 2, ..., N$$

in which case the metric perturbation generated would be

$$h_{\mu\nu}(\omega, r) = K \sum_{k=1}^{N} f_k^{\mu\nu}(\omega) exp(-i\omega|r - r_k|)/|r - r_k|$$

This would be the gravitational analogue of the electromagnetic problem in which we have N infinitesimal current dipoles all carrying currents and located at a discrete set of spatial points. Now coming back to our pulsar problem, we have

$$T^{\mu\nu}(t,r) = f_1^{\mu\nu}(t)\delta^3(r - r_1(t)) + f_2^{\mu\nu}(t)\delta^3(r - r_2(t))$$

and to calculate the metric perturbations generated by this in the frequency domain, we have to first calculate the temporal Fourier transform of a function of time having the form $f(t)\delta^3(r - r_0(t))$. We can instead work completely in time domain using retarded potentials and this would amount to calculating an integral of the form

$$\int f(t - |r - r'|)\delta^3(r' - r_0(t - |r - r'|))d^3r'/|r - r'|$$

$$= \int f(t')\delta^3(r' - r_0(t'))\delta(t' - t + |r - r'|)d^3r'dt'/|r - r'|$$

$$= \int f(t')\delta(t' - t + |r_0(t') - r'|)dt'/|r - r_0(t')|$$

Thus, more generally, if we have a system of N particles moving along trajectories $r_k(t), k = 1, 2, ..., N$, then the energy-momentum tensor generated by this system has the general form

$$T^{\mu\nu}(t,r) = \sum_k f_k^{\mu\nu}(t)\delta^3(r - r_k(t))$$

and hence the metric perturbations induced by this system of moving particles is given by

$$h_{\mu\nu}(t,r) = \sum_{k=1}^N \int_{-\infty}^{\infty} f_k^{\mu\nu}(t')\delta(t' - t + |r_k(t') - r|)dt'/|r - r_k(t')|$$

$$= \sum_k \frac{f_k^{\mu\nu}(t'_k)}{|r - r_k(t'_k)|(1 - (v_k(t'), r - r_k(t'_k))/|r - r_k(t'_k)|)}$$

where t'_k solves

$$t'_k = t - |r - r_k(t'_k)|$$

To a good degree of approximation, ie, when the particle speeds are much smaller than the speed of light, we have the approximation

$$h_{\mu\nu}(t,r) = \sum_k \frac{f_k^{\mu\nu}(t - |r - r_k(t)|)}{|r - r_k(t)|(1 - (v_k(t), r - r_k(t))/|r - r_k(t)|)}$$

If all the $r_k(t)'s$ are in the vicinity of the origin then we have the further approximation

$$h_{\mu\nu}(t,r) = \sum_k \frac{f_k^{\mu\nu}(t - r + \hat{r}.r_k(t))}{r(1 - (v_k(t), \hat{r}))}$$

Assume now that the velocities are small and that the motions of the particles are all periodic with fundamental frequency ω with the positions $r_k(t)$ being small variations around the origin. Then, we can write a Fourier series

$$f_k^{\mu\nu}(t) = \sum_n f_k^{\mu\nu}[n]exp(i\omega nt)$$

and then

$$h_{\mu\nu}(t,r) = \sum_n h_{\mu\nu}[n,t,\hat{r}]exp(i\omega nt)exp(-i\omega nr)/r$$

where

$$h_{\mu\nu}[n,t,\hat{r}] = f_k^{\mu\nu}[n]exp(i\omega n\hat{r}.r_k(t))$$

It is clear that the average energy flux (energy per unit time per unit area in the direction $\hat{r}$) radiated out in the radial direction by this system in the form of gravitational waves has the form

$$\sum_n Re[C(\mu\nu\alpha\beta)\omega^2 n^2 h_{\mu\nu}[n,t,\hat{r}]h_{\alpha\beta}[-n,t,\hat{r}]]$$

Estimation problems in gravitational radiation: It is evident that the source of gravitational waves as elucidated above, depends upon the functions $f_k^{\mu\nu}(t)$ and the positions $r_k(t), k = 1, 2, ..., N$ and in order to know about this source, we must assume that these functions depend upon an unknown parameter vector θ that must be estimated by say allowing the gravitational wave to be incident upon an electromagnetic field and then measuring the change in the pattern of the electromagnetic field. Thus, we write

$$f_k^{\mu\nu}(t) = f_k^{\mu\nu}(t|\theta) = f_{k0}^{\mu\nu}(t) + \sum_s \theta(s)f_{ks}^{\mu\nu}(t)$$

and

$$r_k(t) = r_{k0}(t) + \sum_s \theta(s)r_{k,s}(t)$$

where $\theta(s)$ are small parameters. This assumption of the $\theta(s)'s$ being small parameters amounts to requiring that the uncertainty in the structure of the source generating the gravitational waves is small.

Construction of gravitational wave detectors: Assume that a laser generates an electromagnetic field $F_{\mu\nu}^{(0)}$. In the presence of a gravitational wave $h_{\mu\nu}$, thus field will get perturbed from $F_{\mu\nu}^{(0)}$ to

$$F_{\mu\nu} = F_{\mu\nu}^{(0)} + F_{\mu\nu}^{(1)}$$

where

$$(F^{\mu\nu}\sqrt{-g})_{,\nu} = 0$$

or equivalently,

$$(g^{\mu\alpha}g^{\nu\beta}\sqrt{-g}F_{\alpha\beta})_{,\nu} = 0$$

or equivalently, upto linear orders in the metric perturbations,

$$[(\eta_{\mu\alpha} - h^{\mu\alpha})(\eta_{\nu\beta} - h^{\nu\beta})(1 + h/2)F_{\alpha\beta}]_{,\nu} = 0$$

or

$$[\eta_{\mu\alpha}\eta_{\nu\beta}F^{(1)}_{\alpha\beta}]_{,\nu}$$
$$= [(\eta_{\mu\alpha}h^{\nu\beta} + \eta_{\nu\beta}h^{\mu\alpha} - (1/2)h\eta_{\mu\alpha}\eta_{\nu\beta})F^{(0)}_{\alpha\beta}]_{,\nu}$$

Note that the unperturbed field satisfies the Maxwell equations in flat space-time:

$$[\eta_{\mu\alpha}\eta_{\nu\beta}F^{(0)}_{\alpha\beta}]_{,\nu} = 0$$

Writing

$$A_\mu = A^{(0)}_\mu + A^{(1)}_\mu$$

so that

$$F^{(0)}_{\mu\nu} = A^{(0)}_{\nu,\mu} + A^{(0)}_{\mu,\nu},$$
$$F^{(1)}_{\mu\nu} = A^{(1)}_{\nu,\mu} - A^{(1)}_{\mu,\nu}$$

we have on applying the Lorentz gauge conditions

$$(A^\mu\sqrt{-g})_{,\mu} = 0$$

that

$$[(\eta_{\mu\alpha} - h^{\mu\alpha})(1 + h/2)A_\alpha]_{,\mu} = 0$$

or equivalently,

$$[\eta_{\mu\alpha}A^{(1)}_\alpha]_{,\mu} = [(h^{\mu\alpha} - h\eta_{\mu\alpha}/2)A^{(0)}_\alpha]_{,\mu}$$

6.6   Measuring the gravitational radiation using quantum mechanical receivers

The gravitational wave has the form

$$h_{\mu\nu}(x|\theta) = f_{\mu\nu}(x)^0 + \sum \theta[k]f^k_{x)\mu\nu}$$

where the $\theta[k]s'$ are the unknown parameters to be estimated based on excitations of a quantum mechanical system, say an electron bound to its nucleus described by Dirac's relativistic wave equation. The Dirac equation in a gravitational field is given by

$$[\gamma^a e^\mu_a(x)(i\partial_\mu + \Gamma_\mu(x)) - m]\psi(x) = 0$$

where the spinor connection of the gravitational field is given by

$$\Gamma_\mu(x) = (1/2)[\gamma^a, \gamma^b]e_a^\nu e_{b\nu:\mu}$$

This formula can be derived by noting that the spinor representation of the Lorentz group has generators $(1/4)[\gamma^a, \gamma^b] = \gamma^{ab}$ and hence the spinor connection should have the form

$$\Gamma_\mu = \omega_\mu^{ab} \gamma_{ab}$$

Now the covariant derivative of the tetrad e_μ^a should vanish when the covariant derivative acts on both the tetrad index a and the vector index μ, on the former, via the spinor connection and on the latter via the Christoffel connection. This means that

$$0 = D_\mu e_\nu^a = e_{\nu,\mu}^a - \Gamma_{\mu\nu}^\rho e_\rho^a$$
$$+\omega_\mu^{ab} e_{b\nu}$$

By solving this equation, we get

$$\omega_\mu^{ab} = -e^{b\nu}(e_{\nu,\mu}^a - \Gamma_{\mu\nu}^\rho e_\rho^a)$$

$$= -e^{b\nu} e_{\nu:\mu}^a$$

which is the desired form of the spinor connection of the gravitational field. The curvature of the spinor connection of the gravitational field is given by

$$[\partial_\mu + \Gamma_\mu, \partial_\nu + \Gamma_\nu] =$$

$$\Gamma_{\nu,\mu} - \Gamma_{\mu,\nu} + [\Gamma_\mu, \Gamma_\nu]$$
$$(\omega_{\nu,\mu}^{ab} - \omega_{\mu,\nu}^{ab}) + [\omega_\mu, \omega_\nu]^{ab})\gamma_{ab}$$

where

$$[\omega_\mu, \omega_\nu]^{ab}|\gamma_{ab} = \omega_\mu^{mn} \omega_\nu^{pq}[\gamma_{mn}, \gamma_{pq}]$$

where we have used the notation

$$\gamma_{mn} = [\gamma_m, \gamma_n]$$

When the gravitational wave $h_{\mu\nu}$ is weak, we can write

$$e_{a\mu} = \eta_{a\mu} + h_{a\mu}/2$$

for then

$$\eta_{ab} e_{a\mu} e_{b\nu} = \eta_{\mu\nu} + \eta_{ab}\eta_{a\mu} h_{b\nu}/2 + \eta_{ab}\eta_{b\nu} h_{a\mu} + O(h^2)$$
$$= \eta_{\mu\nu} + h_{\mu\nu} + O(h^2)$$

Equivalently, we can write

$$e_\mu^a = \delta_\mu^a + h_\mu^a/2$$

We then find that upto linear orders in the $h_{\mu\nu}$,

$$\omega_\mu^{ab} = e^{a\nu} e_{\nu:\mu}^b =$$

$$e^{a\nu}(e^b_{\nu,\mu} - \Gamma^\rho_{\nu\mu}e^b_\rho)$$

$$= \eta_{a\nu}h^b_{\nu,\mu}/2 - \eta_{a\nu}\Gamma^b_{\nu\mu}$$

$$= \eta_{a\nu}h^b_{\nu,\mu}/2 - \eta_{a\nu}\eta_{b\rho}(h_{\rho\nu,\mu} + h_{\rho\mu,\nu} - h_{\mu\nu,\rho})/2$$

$$= \eta_{a\nu}\eta_{b\rho}(h_{\mu\nu,\rho} - h_{\rho\mu,\nu})/2$$

Now the Dirac equation in the presence of this background electromagnetic wave can be expressed upto linear orders in the $h_{\mu\nu}$ as

$$0 = [\gamma^a(\delta^\mu_a + h^\mu_a/2)(i\partial_\mu + i\Gamma_\mu) - m]\psi$$

$$= [\gamma^\mu(i\partial_\mu + i\Gamma_\mu) - m]\psi$$

$$+(1/2)\gamma^a h^\mu_a i\partial_\mu\psi$$

$$= [i\gamma^\mu\partial_\mu - m]\psi$$

$$+[i\gamma^\mu\Gamma_\mu + (i/2)\gamma^a h^\mu_a \partial_\mu]\psi$$

where

$$\Gamma_\mu =$$

$$\gamma_{ab}\eta_{a\nu}\eta_{b\rho}(h_{\mu\nu,\rho} - h_{\rho\mu,\nu})/2$$

$$= \gamma^{\nu\rho}(h_{\mu\nu,\rho} - h_{\mu\rho,\nu})/2$$

The operator

$$L = [i\gamma^\mu\Gamma_\mu + (i/2)\gamma^a h^\mu_a \partial_\mu]$$

is a "small perturbation" and hence we can obtain an approximate solution to this gravitationally perturbed Dirac equation using first order perturbation theory. We write the solution as

$$\psi(x) = \psi_0(x) + \psi_1(x)$$

where ψ_0 satisfies the unperturbed Dirac equation:

$$[i\gamma^\mu\partial_\mu - m]\psi_0(x) = 0$$

and $\psi_1(x)$ is the first order perturbation arising from the gravitational perturbation operator L:

$$[i\gamma^\mu\partial_\mu - m]\psi_1(x) + L\psi_0(x) = 0$$

so that if $S(x)$ denotes the electron propgagator:

$$S(x) = \int [\gamma^\mu p_\mu - m + i0]^{-1}exp(-ip.x)d^4p/(2\pi)^4$$

then

$$\psi_1(x) = -\int S(x - y)L\psi_0(y)d^4y$$

Recall that
$$L =$$
$$i[\gamma^{\nu\rho}\gamma^{\mu}(h_{\mu\nu,\rho} - h_{\mu\rho,\nu})/2 + (1/2)\gamma^{a}h_{a}^{\mu}\partial_{\mu}]$$

In the quantum theory of gravity, $h_{\mu\nu}$ as a free gravitational wave is represented by a superposition of graviton creation and annihilation operators having spin two. However, when the gravitational wave is generated by a matter source, then the matter source must also be quantized, for example, if the matter source is a collection of electrons and positrons, then its energy momentum tensor must be that of a second quantized Dirac field and this tensor will therefore be a quadratic form in the electron-positron creation and annihilation operators fields. Thus, in all a gravitational wave coming from a matter source of electrons and positrons will be a linear superposition of free graviton creation and annihilation operators plus a quadratic combination of the electron-positron creation and annihilation operators.

Chapter 7

Some More Concepts and Results in Quantum Information Theory

7.1 Fidelity between two states ρ, σ

Let
$$|x> = \sum_a |x_a \otimes u_a >$$
be a purification of ρ where
$$< u_b | u_a > = \delta(b, a)$$
Let
$$|y> = \sum_a |y_a \otimes v_a >$$
be a purification of σ where
$$< v_b | v_a > = \delta(b, a)$$
Then
$$\rho = \sum_a |x_a >< x_a|, \sigma = \sum_a |y_a >< y_a|$$
Define the matrices
$$X = \sum_a |u_a >< x_a|, Y = \sum_b |v_a >< y_a|$$
Then,
$$X^* X = \rho, Y * Y = \sigma,$$

67

$$X^*Y = \sum <x_a|y_b><u_a|v_b> = <x|y>$$

Then

$$F(\rho, \sigma) = max| <x|y> |$$

where the maximum is taken over all purifications of $|x>$ of ρ and $|y>$ of σ or equivalently over all onb's $\{|u_a>\}$ and $\{|v_b>\}$ of the reference Hilbert space. Now write down the polar decompositions of X and Y as

$$X = U|X| = U\sqrt{\rho}, Y = V|Y| = V\sqrt{\sigma}$$

where U, V are unitary matrices. Then, clearly, from the above equations,

$$F(\rho, \sigma) = max_{U,V}|Tr(X^*Y)| = max_{U,V}||Tr(U\sqrt{\rho}.\sqrt{\sigma}V^*)|$$

$$= max_{U,V}|Tr(\sqrt{\rho}\sqrt{\sigma}V^*U)| = Tr(|\sqrt{\rho}\sqrt{\sigma}|)$$

(where the maximum is taken over all unitaries U, V).

7.2   An identity regarding fidelity

Let $\{E_a\}$ and $\{F_b\}$ be the Choi-Kraus representations of two quantum operations K, K'. We wish to show that there exists a unitary U such that for any two states ρ, σ, we have

$$F(\rho, KoK') \leq F(\rho, KoK_U)$$

where we define

$$F(\rho, K) = <x| < (K \otimes I_R)(|x><x|)|x>$$

with $|x>$ being any purification of ρ with reference Hilbert space $\mathcal{H}_R$. Note that the above expression for $F(\rho, K)$ does not depend upon the purification $|x>$ of the state ρ that has been chosen. To see this, we first observe that any purification $|x>$ of ρ can be expressed as

$$|x> = \sum_a |x_a \otimes e_a>$$

where $\{|e_a>\}$ is an onb for the reference Hilbert space $\mathcal{H}_R$. Then,

$$\rho = Tr_R(|x><x|) = \sum_a |x_a><x_a|$$

and we have

$$(K \otimes I_R)(|x><x|) = \sum_{a,b} K(|x_a><x_b|) \otimes |e_a><e_b|$$

so that

$$< x|(K \otimes I_R)(|x >< x|)|x >= \sum_{a,b} < x_a|K(|x_a >< x_b|)|x_b >$$

$$= \sum_{a,b,c} < x_a|E_c|x_a >< x_b|E_c^*|x_b >$$

$$= \sum_c |\sum_a < x_a|E_c|x_a > |^2 = \sum_c |Tr(E_c \sum_a |x_a >< x_a|)|^2$$

$$= \sum_c |Tr(\rho E_c)|^2$$

which confirms our claim.

7.3 Adaptive probability density tracking using the quantum master equation

$$dU(t) = (-(iH(t) + LL^*/2)dt + LdA(t) - L^*dA(t)^*)U(t)$$

where

$$H(t) = H_0 + f(t)V$$

$f(t)$ is a real valued control forcing function. We wish that $|\psi(t) >= U(t)|\psi_0 \otimes \phi(0) >$ is such that

$$\rho(t) = Tr_2(|\psi(t) >< \psi(t)|)$$

is such that it tracks a desired state $\rho_d(t)$ in the system space. For this, we observe that $\rho(t)$ satisfies

$$\rho'(t) = -i[H(t), \rho(t)] - (1/2)\theta(\rho(t))$$

Then in discrete time form,

$$\rho(t + h) = \rho(t) - ih[H_0 + f(t)V, \rho(t)] - (h/2)\theta(\rho(t))$$

so that

$$\frac{\partial \rho(t + h)}{\partial f(t)} = -ih[V, \rho(t)]$$

and we get using the adaptive LMS algorithm

$$f(t + h) = f(t) - \mu \frac{\partial \| \rho_d(t) - \rho(t) \|^2}{\partial f(t)}$$

$$= f(t) + 2\mu.h.Re(-iTr((\rho_d(t) - \rho(t))[V, \rho(t)])$$

$$= f(t) + 2\mu.h.Im(Tr(\rho_d(t) - \rho(t))[V, \rho(t)])$$

or taking the limit $h \to 0$, we get

$$df(t)/dt = 2\mu.Im(Tr(\rho_d(t) - \rho(t))[V, \rho(t)])$$

Now we can ask a more practical question: Choose a representation of the density operator $\rho(t)$ with respect to the spectral representation of an observable Q, so that

$$< Q|\rho(t)|Q' >= \rho(t, Q, Q')$$

Then, update the control force $f(t)$ so that the diagonal slice $\rho(t, Q, Q)$ tracks a given probability density $p_0(t, Q)$. This means that we should use the adaptive law

$$df(t)/dt = 2\mu.Im \int (p_0(t, Q) - \rho(t, Q, Q)).[V, \rho(t)](Q, Q)dQ$$

$$= -\mu.\frac{\partial}{\partial f(t)} \int (p_0(t, Q) - \rho(t, Q, Q))^2 dQ$$

7.4 Quantum neural networks based on superstring theory

Consider the superstring Hamiltonian expressed in the domain of creation and annihilation operators of the Bosonic and Fermionic fields after truncation to a finite number of terms:

$$L_0 = \sum_{n=1}^{N} a(-n).a(n) + \sum_{n=1}^{N} nS_{-n}.S_n$$

where

$$a(-n).a(n) = \eta_{\mu\nu}a^{\mu}(-n)a^{\nu}(n), S_{-n}.S_n = S_{-n}^a.S_n^a$$

the sum being over repeated indices $\mu, \nu = 0, 1, ..., D-1$ and $a = 1, 2, ..., A$. The truncated superstring propagator is given by

$$\Delta = (L_0 - 1)^{-1} = \int_0^1 z^{L_0-2}dz$$

Note that the CCR and the CAR are

$$[a^{\mu}(n), a^{\nu}(m)] = \eta^{\mu\nu}\delta(n+m), \{S_n^a, S_m^b\} = \delta(a, b)\delta(n+m)$$

Consider a Bosonic-Fermionic vertex function

$$V(k, \sigma) = F(X(0, \sigma), \psi(0, \sigma), \bar{\psi}(0, \sigma))exp(ik.X(0, \sigma))$$

where

$$X(\tau, \sigma) = i \sum_{n=-N, n\neq 0}^{N} (a^{\mu}(n)/n)exp(in(\tau - \sigma))$$

is the left moving Bosonic string and

$$\psi^a(\tau, \sigma) = \sum_{n=-N}^{N} S_n^a exp(in(\tau - \sigma))$$

is the left moving Fermionic string. Note that the Bosonic Lagrangian density is

$$L_B = (1/2)\partial_\alpha X^\mu \partial^\alpha X_\mu,$$

while the Fermionic Lagrangian density is

$$L_F = i\bar{\psi}^a \rho^\alpha \partial_\alpha \psi^a$$

We can choose our function F in the vertex function in various ways that yield after computing the appropriate quantum mechanical elements by interlacing these vertex functions with the propagator quantities that are multilinear forms in the D-vector k^μ and polarization vectors ζ^a that closely resemble those obtained from Feynman diagrammatic calculations applied to the Yang-Mills and super-gravity theories but with some extra string theoretic corrections thereby justifying our hypothesis that quantum field theories are low energy substitutes to string theory. For example, we can choose

$$F = (c_1 k_\mu \partial_\tau X^\mu + c_2 \zeta^a \partial_\tau \psi^a + c_3 k_\mu \partial_\tau X^\mu . \zeta^a \partial_\tau \psi^a) exp(ik.X)$$

Specifically, F is an operator dependent upon the world sheet coordinates of the superstring (τ, σ) and built out of the Bosonic and Fermionic superstring operators in such a way that it represents an interaction between the string at different points along it with other strings attached to it. For example, given such operators $F_1, ..., F_m$, we can construct quantum mechanical probability amplitudes as

$$< \phi_2 | F_1 \Delta . F_2 . \Delta ... \Delta F_m | \phi_1 >$$

This formula is very similar to the situation that we have in quantum field theory in which amplitudes are calculated using the Dyson series expansion that represent the interaction of a system Hamiltonian with an external potential and our matrix elements in the Dyson expansion have the form

$$\int_{-\infty < \tau_1 < ... < \tau_n < \infty} < \phi_2 | U(\infty, \tau_n) V(\tau_n) U(\tau_n, \tau_{n-1}) V(\tau_{n-1}) ... U(\tau_2, \tau_1)$$
$$V(\tau_1) U(\tau_1, -\infty) | \phi_1 > d\tau_1 ... d\tau_n$$

where

$$U(\tau_2, \tau_1) = exp(-i(\tau_2 - \tau_1)H_0)$$

with H_0 being the unperturbed Hamiltonian and $V(\tau)$ the interaction potential. if we have an operator $F(w)$, then

$$\Delta F(w) \Delta = (\int_0^1 z_1^{L_0-2} dz) F(w).(\int_0^1 z_2^{L_0-2} dz_2)$$

$$= \int_0^1 \int_0^1 z_1^{L_0-2} F(w) z_2^{L_0-2} dz_1 dz_2$$

and using the Heisenberg evolution formula for observables, we can write

$$z^{L_0} F(w) z^{-L_0} = F(zw)$$

so

$$z_1^{L_0} F(w) z_2^{L_0} = z_1^{L_0} F(w) z_1^{-L_0} (z_1 z_2)^{L_0}$$
$$= F(z_1 w)(z_1 z_2)^{L_0}$$

or equivalently,

$$z_1^{L_0} F(w) z_2^{L_0} = (z_1 z_2)^{L_0} z_2^{-L_0} F(w) z_2^{L_0}$$
$$= (z_1 z_2)^{L_0} F(w/z_2)$$

These formulas can be used to simplify the computation of the quantum mechanical amplitudes.

A remark: Suppose we apply the potential $V(t)$ only upto time T. Then the above integral for the matrix element in the Dyson series becomes

$$\int_{-\infty < \tau_1 < \dots < \tau_1 < T} < \phi_2 | U(\infty, \tau_n) V(\tau_n) U(\tau_n, \tau_{n-1}) V(\tau_{n-1}) \dots U(\tau_2, \tau_1)$$
$$V(\tau_1) U(\tau_1, -\infty) | \phi_1 > d\tau_1 \dots d\tau_n$$

and after time discretization, we can control the potential $V(T)$ so that the amplitude of transition results in a probability at time T of appropriate form. However, in quantum field theoretic calculations based on the Feynman diagrams, the time parameter does not enter into the picture. This is because, we work in the interaction picture so that $V(t)$ gets replaced by $\tilde{V}(t) = U_0(t)^* V(t) U(t)$ and the corresponding amplitude of scattering by

$$< \phi_2 | \int_{-\infty < t_1 < \dots < t_n < \infty} \tilde{V}(t_n) \dots \tilde{V}(t_1) dt_1 \dots dt_n | \phi_1 >$$

and each $V(t)$ is itself a spatial integral. Thus, the above amplitude assumes the form

$$< \phi_2 | T(\int H_1(x_1) \dots H_1(x_n) d^4 x_1 \dots d^4 x_n) | \phi_1 >$$

where T is the time ordering operator and H_1 is the interaction Hamiltonian density in the interaction picture. After evaluating this matrix element, we end up with a result of the form a sum over products of terms of the form (all in the momentum domain)

$$\int \bar{u}(\phi_2) \Delta_1(f_1 \phi_1, \phi_2, \phi)) \Gamma_1(\phi_1, \phi_2, \phi) \Delta_2(f_2(\phi_1, \phi_2, \phi)) \Gamma_2(\phi_1, \phi_2, \phi) \dots$$
$$\Delta_n(f_n(\phi_1, \phi_2, \phi)) u(\phi_1) d\phi$$

where the Γ_j are appropriate vertex functions. For example in the case of quantum electrodynamics, these vertex functions are corrections to the Dirac Gamma matrices arising from the interaction between electrons, positrons and photons. The $\Delta'_k s$ are propagators of the various particles evaluated at momenta that are some linear combinations f_k of the initial, final and intermediate momenta ϕ_1, ϕ_2, ϕ of the particles. It is precisely because of this final form of the scattering amplitude obtained in quantum field theory and more specifically in quantum electrodynamics, that we look for a similar form for describing scattering processes in superstring theory. For example, consider the Dyson-Schwinger equations that described the interactions between electrons, positrons and photons. These equations are derived from the equations of motion

$$[i\gamma^\mu \partial_\mu - m]\psi(x) = -ieA_\mu(x)\gamma^\mu\psi(x),$$

$$\Box A_\mu(x) = -e\bar{\psi}(x)\gamma_\mu\psi(x)$$

7.5 Designing a quantum neural network for tracking a multivariate pdf based on perturbing a multidimensional harmonic oscillator Hamiltonian by an an-harmonic potential

$$H(t) = H_0 + \sum_{k=1}^{d} f_k(t)V_k$$

where

$$H_0 = \sum_{k=1}^{M} \omega(k)a(k)^*a(k), V_k = g_k(a, a^*), a = (a(k) : k = 1, 2, ..., M),$$

$$a^* = (a(k)^* : k = 1, 2, ..., M)$$

The eigenstates of H_0 are

$$|\mathbf{n}>, \mathbf{n} \in \mathbb{Z}_+^M, \mathbb{Z}_+ = \{0, 1, 2, ...\}$$

and the unperturbed energy eigenvalues are

$$H_0|\mathbf{n}> = \omega.\mathbf{n}|\mathbf{n}>, \omega.\mathbf{n} = \sum_{k=1}^{M} \omega(k)n(k)$$

The perturbed problem has solution $|\psi(t)>$ for its pure state which can be expanded as

$$0|\psi(t)> = \int \psi_0(t, u)|e(u) > d^M u$$

We have assuming that in the expression $g_k(a, a^*)$, all the $a's$ are to the left of all the a^*, that

$$g_k(a, a^*)|e(u)> = g_k(a, \partial/\partial u)|e(u)> = h_k(\partial/\partial u, u)|e(u) >$$

where $h_k(v, u)$ is obtained from $g_k(u, v)$ by pushing all the $v's$ to the left of all the $u's$ in each of its monomoials. Then

$$g_k(a, a^*)|e(u) >= \int \psi_0(t, u) h_k(\partial/\partial u, u)|e(u) > du$$

$$= \int (g_k(u, -\partial/\partial u) \psi_0(t, u))|e(u) > du$$

Further,

$$a(k)^* a(k)|e(u) >= u(k) a(k)^*|e(u) >= u(k)(\partial/\partial u(k))|e(u) >$$

and hence, Schrodinger's equation can equivalently be expressed as

$$i\partial_t \psi(t, u) = \sum_{k=1}^{M} \omega(k) u(k)(\partial \psi(t, u)/\partial u(k)) + \sum_{k=1}^{d} f_k(t) g_k(u, -\partial/\partial u) \psi(t, u)$$

Further, the wave function in position space corresponding to the wave function $\psi(t, u)$ in coherent vector space is given by

$$\psi(t, q) = \int \psi(t, u) < q|e(u) > du$$

where

$$< q|e(u) >= \sum_{\mathbf{n}} (u^{\mathbf{n}}/\sqrt{\mathbf{n}!}) < q|\mathbf{n} >= \sum_{\mathbf{n}} (u^{\mathbf{n}}/\sqrt{\mathbf{n}!}) H_{\mathbf{n}}(q) exp(-|q|^2/2)$$

Our aim is to cause $|\psi(t, q)|^2$ to track a given pdf $p_0(t, q)$ with time.

Now consider the HPQSDE

$$dU(t) = ((-iH + P)dt + LdA - L^* dA^*)U(t), P = LL^*/2$$

The wave function of system $\otimes$ bath is given by

$$\psi(t) = U(t)|\psi(0) >$$

Writing

$$\psi(t) = \int \psi(t, u) \otimes |e(u) > du$$

where $\psi(t, u)$ is a vector in system space while $|e(u) >$ is a coherent vector in Boson Fock space of the noisy bath, we get

$$LdA(t)|\psi(t) >= dt \int (L|\psi(t, u) >)u(t) \otimes |e(u) > du,$$

Thus,

$$< g \otimes e(v)|LdA(t)|\psi(t) >= dt \int < g|L|\psi(t, u) >< e(v)|e(u) > du,$$

$$< g \otimes e(v)|L^*dA(t)^*|\psi(t) >= \bar{v}(t) \int < g|L^*\psi(t,u) >< e(v)|e(u) > du$$

and hence the HPQSDE can also be expressed as

$$id/dt \int < g|\psi(t,u) >< e(v)|e(u) > du =$$

$$\int [- < g|((iH+P)|\psi(t,u) >$$

$$+ < g|L\psi(t,u) > u(t) - \bar{v}(t) < g|L^*\psi(t,u) >] < e(v)|e(u) > du$$

from which we deduce the functional partial differential equation

$$id/dt < g|\psi(t,u) >= - < g|iH+P|\psi(t,u) >$$

$$+u(t) < g|L\psi(t,u) > -(\delta/\delta u(t)) < g|L^*\psi(t,u) >$$

or equivalently,

$$id\psi(t,u)/dt = [(-iH + P) + u(t)L - (\delta/\delta u(t))L^*]\psi(t,u)$$

The aim is to cause the functional pdf $| < q|\psi(t) > |^2$ to track a given functional pdf. Here $q = (q(t) : t \geq 0)$ is the position trajectory. Note that

$$< q|\psi(t) >= \int \psi(t,u) < q|e(u) > du$$

The position field is defined by

$$q(u) = a(u) + a(u)^*$$

and hence its action on a coherent state is given by

$$< e(w)|q(v)|e(u) >=< v|u >< e(w)|e(u) > + < w|u >< e(w)|e(u) >$$

so we can formally write

$$q|e(u) >= (u + \delta/\delta u)|e(u) >$$

or even at a particular time,

$$q(t)|e(u) >= (u(t) + \delta/\delta u(t))|e(u) >$$

A 3-D charged harmonic oscillator interacting with a spatially constant time varying magnetic field directed along the z-axis. The Hamiltonian is

$$H(t) = (1/2)((p_x - eB_0(t)y/2)^2 + (p_y + eB_0(t)x/2)^2) + p_z^2/2 + \omega_0^2(x^2 + y^2 + z^2)/2$$

$$= (p_x^2 + p_y^2 + p_z^2)/2 + eB_0(t)L_z + (\omega(t)^2/2)(x^2 + y^2) + \omega_0^2 z^2/2$$

where

$$\omega(t)^2 = \omega_0^2 + e^2 B_0(t)^2/2$$

This Hamiltonian can be expressed as

$$H(t) = H_0(t) + eB_0(t)L_z$$

with

$$H_0(t) = (p_x^2 + p_y^2 + p_z^2)/2 + (\omega(t)^2/2)(x^2 + y^2) + \omega_0^2 z^2/2$$

commuting with L_z.

7.6   Applied Linear Algebra

Some notions from the representation theory of semisimple Lie algebras.

    1. $Let H_i, X_i, Y_i, i = 1, 2, ..., l$ be the standard generators of the Lie algebra. We have

$$[H_i, H_j] = 0, [H_i, X_j] = a(j,i)X_j, [H_i, X_j] = -a(j,i)Y_j, [X_i, Y_j] = \delta(i,j)H_i$$

Let π be a representation of $\mathfrak{g}$ or equivalently of the universal enveloping algebra $\mathfrak{G}$ in a vector space V. Let $\mathfrak{h} = span\{H_i : 1 \le i \le l\}$ denote the Cartan subalgebra of $\mathfrak{g}$. Let $\lambda \in \mathfrak{h}^*$ and let v be a cyclic vector for π with weight λ. Assume that $\pi(X_i)v = 0, i = 1, 2, ..., l$. Then, it is clear that

$$\pi(\mathfrak{N}_-)v = \mathfrak{g}$$

where $\mathfrak{N}_-$ is the subalgebra of $\mathfrak{G}$ generated by $\{Y_1, ..., Y_l\}$. Indeed write

$$v(i_1, ..., i_k) = \pi(Y_{i_1}...Y_{i_k})v$$

Then $v(i_1, ..., i_k)$ is a weight vector for π with weight $\lambda - \sum_{r=1}^{k} \alpha_{i_r}$ where

$$\alpha_j(H_i) = a(j,i)$$

and moreover

$$\pi(H)v(i_1, ..., i_k) = (\lambda(H) - \sum_{r=1}^{k} \alpha_{i_r}(H))v(i_1, ..., i_k) \in \pi(\mathfrak{N}_-)v, H \in \mathfrak{h}$$

and

$$\pi(X_j)v(i_1, ..., i_k) = ([\pi(X_j), \pi(Y_{i_1})] + \pi(Y_{i_1})\pi(X_j))v(i_2, ..., i_k)$$

$$= (\delta(j, i_1)\pi(H_{i_1}) + \pi(Y_{i_1}\pi(X_j))v(i_2, ..., i_k)$$

$$= \delta(j, i_1)(\lambda(H_{i_1}) - \sum_{r=2}^{k} a(i_r, i_1))v(i_2, ..., i_k)$$

$$+\pi(Y_{i_1})\pi(X_j)v(i_2, ...i_k)$$

and hence by induction, it follows that

$$\pi(X_j))v(i_1, ..., i_k) \in \pi(\mathfrak{N}_-)v$$

In particular, the above arguments show that

$$V = \bigoplus_{1 \le i_1, ..., i_k \le l, k = 1, 2, ..} V_{\lambda - \sum_{r=1}^{k} \alpha_{i_r}}$$

and hence in particular,

$$V_\lambda = \mathbb{C}.v$$

so that
$$dim V_\lambda = 1$$

Now suppose
$$U = \{u \in V : \pi(X_i)u = 0, 1 \leq i \leq l\}$$

Then, we claim that
$$U = \mathbb{C}.v = V_\lambda$$

Let Δ be the set of all roots of $\mathfrak{g}$ and let P denote the set of positive roots relative to a simple system. For $\alpha \in \Delta$, we know that $-\alpha \in \Delta$. Then define $H_\alpha \in \mathfrak{h}$ so that
$$B(H_\alpha, H) = \alpha(H), H \in \mathfrak{h}$$

Then, for $X_\alpha \in \mathfrak{g}_\alpha, X_{-\alpha} \in \mathfrak{g}_{-\alpha}$, we have
$$[X_\alpha, X_{-\alpha}] = \lambda.H_\alpha$$

for some $\lambda \in \mathbb{C}$. Then, since for all $H \in \mathfrak{h}$, we have
$$\lambda\alpha(H) = \lambda.B(H_\alpha, H) =$$
$$B([X_\alpha, X_{-\alpha}], H) = B(X_\alpha, [X_{-\alpha}, H]) =$$
$$\alpha(H)B(X_\alpha, X_{-\alpha})$$

it follows that
$$\lambda = B(X_\alpha, X_{-\alpha})$$

and therefore,
$$[X_\alpha, X_{-\alpha}] = B(X_\alpha, X_{-\alpha})H_\alpha$$

We can choose $X_\alpha, \alpha \in \Delta$ such that
$$B(X_\alpha, X_{-\alpha}) = 2/ < \alpha, \alpha >$$

where
$$< \alpha, \beta > = < H_\alpha, H_\beta > = \alpha(H_\beta)$$

Here $< X, Y > = B(X, Y)$. Now define
$$\bar{H}_\alpha = 2H_\alpha/ < \alpha, \alpha >$$

and then we get
$$\alpha(\bar{H}_\alpha) = 2, \beta(\bar{H}_\alpha) = 2 < \beta, \alpha > / < \alpha, \alpha >$$

Representations of $\mathfrak{sl}(2, \mathbb{C})$. The standard generators of $\mathfrak{sl}(2, \mathbb{C})$ are (H, X, Y) satisfying the commutation relations
$$[H, X] = 2X, [H, Y] = -2Y, [X, Y] = H$$

Let $sl(2, \mathbb{C})$ act on a finite dimensional vector space V via an irreducible representation ρ, ie, for $v \in V$ and $Z \in sl(2, \mathbb{C})$,

$$Z.v = \rho(Z)v$$

Let λ be an eigenvalue of $\rho(H)$ with eigenvector v. Then $\rho(X)v$ is either zero or else an eigenvector of $\rho(H)$ with eigenvalue $\lambda + 2$ and likewise, $\rho(Y)v$ is either zero or else an eigenvector of $\rho(H)$ with eigenvalue $\lambda - 2$. Since V is finite dimensional and eigenvectors of any matrix corresponding to distinct eigenvalues are linearly independent, it follows from this argument that there exists a nonzero vector v_0 that is an eigenvector of $\rho(H)$ with some eigenvalue λ_0 and also satisfies $\rho(X)v_0 = 0$ and further that there is a finite non-negative integer m such that $\rho(Y)^m v \neq 0$ but $\rho(Y)^{m+1} v = 0$. Then, defining

$$v_j = \rho(Y)^j v_0, j = 0, 1, ..., m$$

it follows from the irreducibility of ρ that $\{v_j : 0 \leq j \leq m\}$ is a basis for V and that

$$\rho(H)v_j = (\lambda_0 - 2j)v_j, 0 \leq j \leq m$$

It is clear that $\rho(Y)v_j = v_{j+1}$ and that $\rho(X)v_j$ is proportional to v_{j-1}. We shall evaluate the proportionality constant. To do so, observe that

$$Xv_j = XY^j v_0 = [X, Y^j]v_0$$

$$= \sum_{r=0}^{j-1} Y^{j-1-r}[X, Y]Y^r v_0 = \sum_{r=0}^{j-1} Y^{j-1-r}HY^r v_0$$

$$= \sum_{r=0}^{j-1} Y^{j-1-r}([H, Y^r] + Y^r H)v_0$$

$$= \sum_{r=0}^{j-1} Y^{j-1-r}(-2rY^r + \lambda_0 Y^r)v_0$$

$$= \sum_{r=0}^{j-1} (\lambda_0 - 2r)Y^{j-1}v_0$$

$$= (\lambda_0 j - j(j-1))v_{j-1}$$

Thus the proportionality constant is $j(\lambda_0 - j + 1)$. Now, we further observe that $Y^{m+1}v_0 = 0$ implies $XY^{m+1}v_0 = 0$ and hence by the above argument,

$$\lambda_0(m+1) - (m+1)m = 0$$

or equivalently,

$$\lambda_0 = m$$

It follows that the eigenvalues of $\rho(H)$ are precisely, $m, m-1, ..., -m$ with the corresponding eigenvectors being $Y^j v_0 = v_j, j = 0, 1, ..., m$. In particular, we have deduced the fundamental result that all the eigenvalues of $\rho(H)$ for any irreducible representation ρ of $sl(2, \mathbb{C})$ are of the form $|j|, j \leq m$ where m is some non-negative integer. For $m = 0$, we get a one dimensional irreducible representation of $sl(2, \mathbb{C})$ that is the trivial identity representation of the group or equivalently the zero representation of the Lie algebra.

Chapter 8

Quantum Field Theory, Quantum Statistics, Gravity, Stochastic Fields and Informationy

8.1 Rate distortion theory for ergodic sources

Let $X_1, X_2, ...$ be iid random variables assuming values in a finite alphabet A in accordance with the probability distribution P_1. For each $n \geq 1$, construct a code $C_n : A^n \to A^n$. We write X for $(X_1, ..., X_n)$ and then $C_n(X) \in A^n$ with components $C_n(X)_i, i = 1, 2, ..., n$. We say that $C_n(X)$ is a code word corresponding to the source word $X = (X_1, ..., X_n)$. Let $C_n(A^n)$ denote the range of the code C_n, ie,

$$C_n(A^n) = \{C_n(x) : x \in A^n\} \subset A^n$$

$|C_n(A^n)|$ is the cardinality of the code C_n and is also called the size of the code. The rate of the code is

$$R_n(C) = log(|C_n(A^n))|)/n$$

Define

$$\rho_n(C) = \mathbb{E}(n^{-1} \sum_{k=1}^{n} \rho(X_k, C_n(X)_k)$$

$\rho_n(C)$ is called the distortion of the code C_n. Further, for any probability distribution Q on $A \times A$, define

$$\rho(Q) = \int_{A \times A} \rho(x, y) dQ(x, y)$$

Here $,\rho$ is a metric on A. In data compression theory, we wish to select a sequence of codes C_n each of as small a size as compared with $|A|^n = |A^n|$ in such a way that $limsup\rho_n(C) \leq D$ where D is a fixed positive real number, called the allowable distortion threshold. Shannon proved the following theorem in this regard: Define the rate distortion function

$$R(D) = inf\{H(Q|Q_1 \times Q_2) : \rho(Q) \leq D, Q_1 = P_1\}$$

where Q_1 and Q_2 are the marginals of Q. The infimum is over all probability distributions Q on $A \times A$. Shannon proved that given $D > 0$, for each $\delta > 0$, there exists a sequence of codes $C_n : A^n \to A^n, n = 1, 2, ...$ such that

$$limsup R_n(C) \leq R(D) + \delta$$

and simultaneously
$$limsup\rho_n(C) \leq D$$

and also the converse, namely that if a sequence of codes $C_n : A^n \to A^n, n \geq 1$ satisfies the condition

$$limsup R_n(C) \leq R(D)$$

then
$$limsup\rho_n(C) > D$$

for any $D > 0$. Heuristically speaking, this means that given a distortion threshold D, a sequence of codes must have a rate of at least $R(D)$ in order for the distortion level of this sequence not to exceed D.

Proof of the direct part based on large deviation theory and Shannon's random coding argument. For a given probability distribution Q on $A \times A$ and for $x \in A^n$, define

$$S_n(x) = \{y \in A^n : n^{-1}\sum_{k=1}^{n}\rho(x_k, y_k) \leq \rho(Q) + \delta\}$$

Define a random code C_n on A^n consisting of the code-words $\{Y(i) = (Y(i,1), ..., Y(i.n)) : i = 1, 2, ..., k_n\}$ where the $Y(i,j), i = 1, 2, ..., k_n, j = 1, 2, ..., n$ are iid r.v's with distribution Q_2, the second marginal of Q independently of X. Here

$$k_n = [exp(n(H(Q|Q_1 \times Q_2) + \delta))]$$

If $C_n \cap S_n(x) \neq \phi$, we pick an element $C_n(x)$ from this intersection. Then, we clearly have from the definition of the set $S_n(x)$ that

$$n^{-1}\sum_{k=1}^{n}\rho(x_k, C_n(x)_k) \leq \rho(Q) + \delta + \rho_{max}\chi_{C_n \cap S_n(x)=\phi} - - - (1)$$

where $C_n = C_n(A^n)$. Now, for fixed $x \in A^n$. But,

$$P(C_n \cap S_n(x) = \phi) = P((Y(i) \notin S_n(x), i = 1, 2, ..., k_n)$$

$$= P(Y(1) \notin S_n(x))^{k_n}$$

$$= (1 - P(Y(1) \in S_n(x)))^{k_n} \leq exp(-k_n P(Y(1) \in S_n(x)))$$

$$\leq exp(-exp(n\delta)) \to 0$$

almost surely for all x distributed as Q_1^n because as we shall presently show,

$$liminf n^{-1} log(P(Y(1) \in S_n(x))) \geq -H(Q|Q_1 \times Q_2)$$

almost surely for all sequences x having the iid Q_1 distribution. From this and (1), it follows immediately on taking expectations after replacing x by X (first w.r.t X and then w.r.t the random code C_n) that

$$limsup \mathbb{E}(\rho_n(C)) \leq \rho(Q) + \delta$$

It follows then that there exists a sequence of non-random codes $C_n^{(0)}$ such that

$$limsum \rho_n(C^{(0)}) \leq \rho(Q) + \delta$$

and further the size of $C_n^{(0)}$ is clearly k_n and therefore

$$limsup_n log(|C_n^{(0)}|)/n = limsup log(k_n)/n = H(Q|Q_1 \times Q_2) + \delta$$

From this, the direct part of Shannon's rate distortion theorem follows at once. Now, we must prove the large deviation result stated during the proof. First define the probability measure Q_λ on $A \times A$ by

$$dQ_\lambda(x,y) = exp(\lambda\rho(x,y))dQ_1(x) \times dQ_2(y) / \int exp(\lambda\rho(x,y'))dQ_2(y')$$

Note that Q_λ has first marginal Q_1. Now,

$$0 \leq H(Q|Q_\lambda)$$

$$= H(Q|Q_1 \times Q_2) - \int dQ_1(x).log(\int exp(\lambda\rho(x,y))dQ_2(y))$$

$$-\lambda\rho(Q) - - - (2)$$

for any $\lambda \in \mathbb{R}$. Define

$$\Lambda_Q(\lambda) = \int dQ_1(x).log(\int exp(\lambda\rho(x,y)dQ_2(y))$$

and let

$$\Lambda_Q^*(\xi) = sup_\lambda(\lambda\xi - \Lambda_Q(\lambda))$$

Then, we get from (2),

$$H(Q|Q_1 \times Q_2) \geq \Lambda_Q^*(\rho(Q))$$

Further, we have

$$P(Y(1) \in S_n(x)) = P(Z_n(x) \le \rho(Q) + \delta)$$

where

$$Z_n(x) = n^{-1} \sum_{k=1}^{n} \rho(x_k, Y(1, k))$$

with the $Y(1, k)'s$ being iid with distribution Q_2. The Gartner-Ellis limiting logarithmic moment generating function for the $Z_n(x)'$ is given by

$$lim n^{-1}.log \mathbb{E} exp(n\lambda Z_n(x))$$

$$= lim n^{-1}] sum_{k=1}^{n} log \int exp(\lambda \rho(x_k, y)) dQ_2(y)$$

$$= \int dQ_1(x).log \int exp(\lambda \rho(x, y)) dQ_2(y) = \Lambda_Q(\lambda)$$

the last equation being an almost sure equation assuming that the sequenc $x = (x_k)$ has the iid $Q_1 = P_1$ distribution. Thus, the large deviation lower bound, we have $a.s. Q_1$

$$lim inf n^{-1} log(P(Y(1) \in S_n(x))) \ge -inf\{\Lambda_Q^*(\xi) : \xi \le \rho(Q) + \delta\}$$

$$\ge -\Lambda_Q(\rho(Q)) \ge -H(Q|Q_1 \times Q_2)$$

which completes the proof of the direct part of Shannon's rate distortion compression theorem.

Remark on the relationship with the zero distortion Shannon noiseless coding theorem. If $D = 0$, ie, zero distortion is required, then the asymptotic rate of compression becomes $H(Q|Q_1 \times Q_2) + \delta$ with $\rho(Q) = 0$. Now, $\rho(Q) = 0$ means that for a given input source alphabet $x \in A$, the output alphabet $y = x$ a.s. Q, so that

$$\int \rho(x, y) dQ(x, y) = 0$$

It in other words,

$$H(Q) = H(Q_1), H(Q_2) = H(Q_1)$$

and therefore

$$H(Q|Q_1 \times Q_2) = H(Q_1) + H(Q_2) - H(Q) = H(Q_1)$$

and the compression rate for zero distortion then reduces to the classical Shannon result $H(Q_1)$.

Converse part: Let C_n be a sequence of codes $(C_n : A^n \to A^n)$ an let $R_n(C) = log(|C_n|)/n$. Suppose

$$\rho_n(C) = \mathbb{E}(n^{-1} \sum_{k=1}^{n} \rho(X_k, C_n(X)_k)) \le \rho(Q)$$

Define any probability distribution Q_n on $A^n \times A^n$ so that its first marginal is $Q_1^n = P_1^n$ and Q_n is concentrated on the sets $(x, C_n(x)), x \in A^n$. Let Q_{2n} denote the second marginal of Q_n. We can write

$$Q_{2n}(y) = \sum_{x \in A^n} f_n(x, y) Q_1^n(x), y \in A^n$$

Thus,

$$Q_n(x, y) = f_n(x, y) Q_1^n(x) Q_{2n}(y)$$

or equivalently,

$$(dQ_n / dQ_1^n \times dQ_{2n})(x, y) = f_n(x, y)$$

Since

$$1 = \sum_y Q_{2n}(y)$$

it follows that

$$0 \le f_n(x, y) Q_1^n(x) \le 1$$

Now, since Q_{2n} is concentrated on the set $C_n(A^n)$, it follows that

$$H(Q_{2n}) \le log|C_n| = nR_n(C)$$

Thus,

$$H(Q_n | Q_1^n \times Q_{2n}) = H(Q_1^n) + H(Q_{2n}) - H(Q_n) \le H(Q_{2n}) \le nR_n(C)$$

Now,

Problem: Let Q be a probability distribution on $A^n \times A^n$ and let Q_X, Q_Y be its two marginals on A^n. Assume that $Q_X = P^n$ is an iid product measure, and let $Q_{Y,i}$ denote the marginals of Q_Y. Note that P and $Q_{Y,i}$ are probability distributions on A. Then, show that

$$H(Q | Q_X \times Q_Y) \ge \sum_{i=1}^n H(Q_i | P_1 \times Q_{2i})$$

where Q_i is the ith marginal of Q on $A \times A$ when we write $A^n \times A^n = (A \times A)^n$.

Solution:

$$H(Q) \le \sum_i H(Q_i), H(Q_Y) \le \sum_i H(Q_{Y,i})$$

Thus

$$H(Q | Q_X \times Q_Y) = -H(Q) + H(Q_X) + H(Q_Y)$$

$$= -H(Q) + nH(P_1) + H(Q_Y) = nH(P_1) - H(X_1, ..., X_n | Y_1, ..., Y_n)$$

On the other hand,

$$\sum_{i=1}^n H(Q_i | P_1 \times Q_{2i})$$

$$= n.H(P_1) - \sum_i (H(X_i, Y_i) - H(Y_i)) = nH(P_1) - \sum_i H(X_i|Y_i)$$

Thus, the problem amounts to showing that

$$H(X_1, ..., X_n|Y_1, ..., Y_n) \le \sum_i H(X_i|Y_i)$$

This follows immediately from the fact that

$$H(X_i|Y_i) \ge H(X_i|Y_1, ..., Y_n) \forall i$$

and therefore

$$\sum_i H(X_i|Y_i) \ge \sum_i H(X_i|Y_1, ..., Y_n) \ge H(X_1, ..., X_n|Y_1, ..., Y_n)$$

Remark: For any r.v's X_1, X_2, Y, we have

$$H(X_1|Y) + H(X_2|Y) \ge H(X_1, X_2|Y)$$

because

$$H(X_1, X_2|Y) = H(X_1|X_2, Y) + H(X_2|Y) \ge H(X_1|Y) + H(X_2|Y)$$

since

$$H(Z|U) - H(Z|U, V) \ge 0$$

Note that

$$H(Z|U) - H(Z|U, V) = - \sum_{z,u,v} p(x, u, v) log(p(z|u)) + \sum_{z,u,v} p(z, u, v) log(p(z|u, v))$$

$$= \sum_{z,u,v} p(z, u, v) log(p(z|u, v)/p(z|u))$$

$$= \sum_{u,v} p(u, v). \sum_z p(z|u, v) log(p(z|u, v)/p(z|u)) \ge 0$$

8.2   Problems

Prove that the function $-x.log(x)$ on the positive reals is concave and hence deduce concavity of the entropy.

   Prove that doubly stochastic matrices of a given dimension form a convex set whose extreme points are the permutation matrices. Deduce that if p is a probability vector and Q a doubly stochastic matrix, then

$$H(Qp) \le H(p)$$

ie, evolution of a Markov chain under a doubly stochastic matrix always decreases the entropy.

8.3 Simulation of time varying joint probability densities using Yang-Mills gauge theories

Vertex function in quantum field theory: Let

$$< 0|T(\psi(x)\bar{\psi}(y)A^{\nu}(z)))|0 >$$

$$= \int S'(p)\Gamma_{\mu}(p, p')S'(p')D^{\mu\nu'}(p-p')exp(i(p.(x-z)-p'.(y-z)))d^4pd^4p' ---(1)$$

Then $\Gamma_{\mu}(p, p')$ is called the vertex function.

Application to the computation of the electron self energy function.

$$[i\gamma.\partial - m]\psi(x) = -e\gamma^{\nu}\psi(x).A_{\nu}(x)$$

Thus, if

$$S'(x - y) =< 0|T(\psi(x)\bar{\psi}(y))|0 >,$$

then

$$[i\gamma.\partial_x - m]S'(x - y) = \gamma^0\delta^4(x - y) - e\gamma^{\nu} < 0|T(\psi(x)\bar{\psi}(y)A_{\nu}(x))|0 >$$

$$= \gamma^0\delta^4(x - y) - e\gamma_{\nu}\int S'(p)\Gamma_{\mu}(p, p')S'(p')D^{\mu\nu'}(p - p')exp(ip'.(x - y))d^4pd^4p'$$

Taking Fourier transform of this equation w.r.t $x - y$ gives

$$[\gamma.p - m]S'(p) = \gamma^0 - e\gamma_{\nu}\int S'(q)\Gamma_{\mu}(q, q')S'(q')D^{\mu\nu'}(q - q')\delta(q' - p)d^4qd^4q'$$

or equivalently,

$$S'(p) = S(p) - e\gamma_{\nu}S(p)\int S'(q)\Gamma_{\mu}(q, p)S'(p)D^{\mu\nu'}(q - p)d^4q$$

$$= S(p) + S(p)\Sigma(p)S'(p)$$

where $\Sigma(p)$ is the electron self-energy function given by

$$\Sigma(p) = -e\gamma_{\nu}\int S'(q)\Gamma_{\mu}(q, p)D^{\mu\nu'}(q - p)d^4q$$

The notation used here is as follows: $D^{\mu\nu}, S$ are respectively the photon and electron propagators in the absence of interactions between the two while $D^{\mu\nu'}$ and S' are respectively the exact photon and electron propagators, ie, after taking into account interactions between the two. Upto first order perturbation theory,

$$\Sigma(p) = -e\gamma_{\nu}\int S(q)\gamma_{\mu}D^{\mu\nu}(q - p)d^4q$$

since in first order perturbation theory, the vertex function is

$$\Gamma_{\mu}(p, p') = \gamma_{\mu}$$

This can be seen as follows:

$$< 0|T(\psi(x)\bar{\psi}(y)A^\nu(z))|0 >=$$

$$e < 0|T(\psi(x)\bar{\psi}(y)(\int D^{\mu\nu}(z-z')\bar{\psi}(z')\gamma_\mu\psi(z')d^4z'))|0 >$$

$$= e\int < 0|T(\psi(x)\bar{\psi}(y)\bar{\psi}(z')\gamma_\mu\psi(z'))|0 > D^{\mu\nu}(z-z')d^4z'$$

In first order perturbation theory, we can evaluate the above time ordered vacuum expectation by using the bare electron propagator to get

$$< 0|T(\psi(x)\bar{\psi}(y)A^\nu(z))|0 >$$

$$= e\int S(x-z')\gamma_\mu S(z'-y)D^{\mu\nu}(z-z')d^4z'$$

and then a comparison with (1) after transforming the integral to the momentum domain using the Fourier transform shows that

$$\Gamma_\mu(p,p') = \gamma_\mu$$

In terms of the vertex function, we have thus calculated the change in the electron propagator caused by radiative effects, ie, by the interactions between the electron-positron field and the photon field. Let us likewise calculate the change in the photon propagator caused by radiative effects in terms of the vertex function. We start with

$$\Box A^\mu(x) = e\bar{\psi}(x)\gamma^\mu\psi(x)$$

and hence if

$$D^{\mu\nu'}(x-y) =< 0|T(A^\mu(x)A^\nu(y))|0 >$$

we get

$$\Box_x D^{\mu\nu'}(x-y) = \eta^{\mu\nu}\delta^4(x-y) + e < 0|T(\bar{\psi}(x)\gamma^\mu\psi(x)A^\nu(y)))|0 >$$

This gives

$$D^{\mu\nu'}(x-y) = D^{\mu\nu}(x-y) + e\int D(x-z) < 0|T(\bar{\psi}(z)\gamma^\mu\psi(z)A^\nu(y))|0 > dz$$

where we recall that

$$\eta^{\mu\nu}D(x-y) = D^{\mu\nu}(x-y)$$

Now,

$$< 0|T(\bar{\psi}(z)\gamma^\mu\psi(z)A^\nu(y))|0 >=$$

$$\gamma^\mu_{ab} < 0|T(\psi_b(z)\bar{\psi}_a(z)A^\nu(y))|0 >$$

$$= Tr(\gamma^\mu < 0|T(\psi(z)\bar{\psi}(z)A^\nu(y))|0 >)$$

$$= Tr(\gamma^\mu \int S'(p)\Gamma_\rho(p,p')S'(p')D^{\rho\nu'}(p-p')exp(ip.(z-y) - ip'.(z-y))d^4pd^4p')$$

This give us the following equation for the corrected photon propagator:

$$D^{\mu\nu'}(x-y) = D^{\mu\nu}(x-y) + eTr(\int D(x-z)\gamma^\mu S'(p)\Gamma_\rho(p,p')S'(p')D^{\rho\nu'}(p-p')$$

$$exp(i(p-p').(z-y))d^4pd^4p'd^4z)$$

$$= D^{\mu\nu}(x-y) + e.Tr(\int D(x-y-z)\gamma^\mu \Gamma_\rho(p,p')D^{\rho\nu'}(p-p')exp(i(p-p').z)d^4pd^4p'd^4z)$$

$$= D^{\mu\nu}(x-y) + eTr\int (D(p-p')\gamma^\mu \Gamma_\rho(p,p')D^{\rho\nu'}(p-p')exp(i(p-p').(x-y))d^4pd^4p'$$

or equivalently in the momentum domain,

$$D^{\mu\nu'}(p) = D^{\mu\nu}(p) + eD(p)Tr(\gamma^\mu \int \Gamma_\rho(p+k,k)d^4k)D^{\rho\nu'}(p)$$

$$= D^{\mu\nu}(p) + eD^{\mu\alpha}(p)Tr(\gamma_\alpha \int \Gamma_\rho(p+k,k)d^4k)D^{\rho\nu'}(p)$$

$$= D^{\mu\nu}(p) + D^{\mu\alpha}(p)\Pi_{\alpha\rho}(p)D^{\rho\nu'}(p)$$

where the polarization tensor $\Pi_{\alpha\rho}(p)$ is given by

$$\Pi_{\alpha\rho}(p) = eTr(\gamma_\alpha \int \Gamma_\rho(p+k,k)d^4k)$$

8.4 An application of the radiatively corrected propagator to quantum neural network theory

Consider the scattering of two electrons having initial four momenta p_1, p_2 and final momenta p'_1, p'_2 with a photon being exchanged during the scattering process. There are two possible Feynman diagrams for this process. According to the first diagram, the scattering amplitude is given by

$$A(p'_1, p'_2|p_1, p_2) = \bar{u}(p'_1)\gamma^\mu u(p_1).\bar{u}(p'_2)\gamma^\nu u(p_2)D'_{\mu\nu}(p_1 - p'_1)\delta^4(p'_1 + p'_2 - p_1 - p_2)$$

Now consider the same scattering amplitude but now taking into account the extra effect that when the photon is being exchanged during the scattering process, it polarizes into an electron and a positron via a single loop. The amplitude for such a process is given by

$$\bar{u}(p'_1)\gamma^\mu u(p_1)D_{\mu\nu}(p_1-p'_1)[\int Tr[S'(q)\gamma_\nu S'(p_1-p'_1-q)\gamma_\rho]d^4q]\bar{u}(p'_2)\gamma^\rho u(p_2)\delta^4$$

$$(p'_1+p'_2-p_1-p_2)D(p'_2-p_2)$$

where $D(p) = 1/(p^2 + i0)$ and $S'(q)$ is the radiatively corrected electron propagator. Upto one loop orders, the total scattering amplitude via the first channel is the sum of these two amplitudes. Now consider an even more complex situation when there is an external c-number control photon vector potential $A_\mu^c(k)$ line carrying four momentum k that is absorbed by the first electron prior to its getting scattered. In this case, the scattering amplitude taking into account this c-number line and the polarization of the exchanged photon into an electron and a positron via a one loop factor is given by

$$A(p_1', p_2'|p_1, p_2, k) =$$

$$A_\beta^c(k)\bar{u}(p_1')\gamma^\mu S'(p_1+k)\gamma^\beta u(p_1)$$

$$.D_{\mu\nu}(p_1-p_1'+k)[\int Tr[S'(q)\gamma_\nu S'(p_1-p_1'+k-q)\gamma_\rho]d^4q]\bar{u}(p_2')\gamma^\rho u(p_2)\delta^4$$

$$(p_1'+p_2'-p_1-p_2-k)D(p_2'-p_2)$$

If however, all the frequencies and wavelengths are present in the external c-number photon line, then to obtain the scattering amplitude, we have to integrate the above expression over all external c-number photon four momenta k to obtain the following expression for the scattering amplitude:

$$A(p_1', p_2'|p_1, p_2) = \int A(p_1', p_2'|p_1, p_2, k)d^4k =$$

$$A_\beta^c(p_1' + p_2' - p_1 - p_2)\bar{u}(p_1')\gamma^\mu S'(p_1' + p_2' - p_2)\gamma^\beta u(p_1)$$

$$.D_{\mu\nu}(p_2' - p_2)[\int Tr[S'(q)\gamma_\nu S'(p_2' - p_2 - q)\gamma_\rho]d^4q]\bar{u}(p_2')\gamma^\rho u(p_2)$$

The scattering amplitude when our c-number potential A_μ^c is applied upto time t is given by $A_t(p_1', p_2'|p_1, p_2)$ which is obained by replacing $A_\beta^c(k)$ by

$$A_\beta^c(t,k) = \int_{\tau \le t, r \in \mathbb{R}^3} A_\beta^c(\tau, r) exp(-i(k^0\tau - K.r))d\tau d^3r$$

We can then control A_β^c at time $t + h$ so that $|A_{t+h}(p_1', p_2'|p_2, p_2)|^2$ is close to some given transition probability density $Q_{t+h}(p_1', p_2'|p_1, p_2)$ using an algorithm like the stochastic gradient algorithm.

The logic underlying this formalism is the Dyson series expansion of the scattering amplitude between two states: If $V(s)$ is the interaction potential in the interaction representation and this potential is applied upto time t and thereafter switched off, then the scattering amplitude between an initial state $|\phi_i>$ and a final state $|\phi_f>$ is given by

$$A_\infty(\phi_f|\phi_i) = \sum_{n \ge 1}(-i)^n < \phi_f| \int_{-\infty < t_n < ... < t_1 < \infty} V(t_1)...V(t_n)dt_1...dt_n|\phi_i >$$

$$= \sum_{n \ge 1}(-i)^n < \phi_f| \int_{-\infty < t_n < ... < t_1 \le t} V(t_1)...V(t_n)dt_1...dt_n|\phi_i >$$

$$= A_t(\phi_f|\phi_i)$$

8.5 An experiment involving the measurement of Newton's gravitational constant G

Using a sensitive torsion balance located underground so that there would be no atmospheric disturbance in the torsion experiment. Two masses were connected to this torsion string and the force between the two masses could be measured by measuring the amount of twist of the torsion string. From measurements of this force, one could in principle calculate G using Newton's inverse square law of gravitation. Now one could try this experiment with other masses made of other materials and again measure G. This process of measuring G with different masses attached, if sensitive enough, would be able to tell us whether G was the same for all materials or not. My boss had in mind a theory called the theory of the fifth force which said that G would not be a constant because the fifth force on a body due to another need not be proportional to their masses, perhaps it may depend on the the composition of the material like the number of neutrinos in the two masses. This conjecture, if proved right, would make one have to revise Einstein's principle of equivalence of the proportionality between inertial and gravitational masses. Specifically, the dynamics of a mass m in the gravitational field of another mass M would have to be expressed as

$$ma = GMm/r^2 + F$$

where F is the fifth force exerted by M on m. Thus, m would not cancel from both sides and hence we would get

$$G = \frac{r^2(ma - F)}{Mm} = \frac{r^2 a}{M} - \frac{r^2 F}{Mm}$$

In the absence of the fifth force, $F = 0$ and one would recover the classical result that a is independent of m and proportional to M provided that G was a constant. The presence of a nonzero F would either make G a constant and a a function of m or else it would make G non-constant and dependent on m thereby causing the proportionality of inertial and gravitational masses to get violated. The conjecture is that the neutrino is its own antiparticle and that a major portion of the unobservable dark matter in the universe was composed of neutrinos which was the source of the fifth force. Model the dynamics of the torsion oscillator taking noise into account and develop a statistical theory for its oscillations. The celebrated "Fluctuation-Dissipation Theorem" developed primarily by Einstein and polished by the Japanese physicist Ryogo Kubo which states that there must be a relationship between the fluctuation coefficient and the dissipation coefficient in a stochastic dynamical system in order that the system be in thermal equilibrium at a given temperature T. From the standpoint of stochastic differential equations, we can formulate this principle as follows: Consider a mechanical system with n canonical coordinates and n canonical momenta. Set up the Hamilton equations of motion and add non-conservative terms to this differential system in the form of dissipative forces proportional to

the canonical momenta and random forces proportional to white noise or equivalently to the time derivative of Brownian motion. Transform this system into a
system of Ito stochastic differential equations and write down the Fokker-Planck
or forward Kolmogorov equation for the evolution of the joint probability density of the canonical coordinates and momenta. Consider the situation when
the Gibbs density $C.exp(-H/kT)$ where H is the Hamiltonian and T the temperature, is an equilibrium solution to this Fokker-Planck equation. One then
recovers a generalization of the famous Einstein fluctuation-dissipation theorem
that relates the dissipation coefficients to the fluctuation /diffusion coefficients.
The idea behind applying this theorem to our torsion oscillator is that we cannot in our model select the dissipation and diffusion coefficients at random, they
must always be selected so that the fluctuation-dissipation theorem is satisfied
and then from the measured dynamics of the resulting "stochastic torsion oscillator" we can hope to get a reliable estimate of its parameters, one of them
being G.

8.6 Extending the fluctuation-dissipation theorem

To a general system of stochastic differential equations with the condition that
a given density be an equilibrium density for this system. Specifically, if one
considers the coupled system of sde's

$$dX(t) = f(X(t))dt + g(X(t))dB(t)$$

then the condition that a function $p(X)$ be an equilibrium density for this system
is that

$$-\nabla_X^T(f(X)p(X)) + (1/2)Tr(\nabla_X \nabla_X^T(gg^T(X)p(X))) = 0$$

This determines a realtionship between the drift coefficient vector valued function $f(X)$ and the diffusion coefficient matrix valued function $g(X)$

8.7 A discrete Poisson collision approach to Brownian motion

The origin of this idea is Einstein's little book on "investigations into the theory
of the Brownian movement" of around sixty odd pages where Einstein had derived from basic collision theory, the diffusion equation for the pdf of Brownian
motion and had even suggested using Stokes' formula for the viscous force on
a spherical pollen particle moving in a liquid, the equation of continuity for a
fluid of particles and the ideal gas equation, a formula for the diffusion coefficient in terms of the temperature, Boltzmann's constant, Avogadro's number,
the viscosity of the fluid and the radius of the pollen particle. Einstein had
then by solving the diffusion equation, related the mean square deviation in

the position of the pollen grain after a given time duration to the diffusion coefficient and hence suggested a method to calculate Avogadro's number from measurements taken on the displacment of several pollen particles undergoing Brownian motion inside a liquid. This experiment, Einstein had said, would provide a concrete proof of the kinetic molecular theory of matter, that matter was indeed composed of atoms and molecules. It was at that time one of the miraculous pieces of theoretical phyisics because it was not clear even after Rutherford's experiment and Bohr's model of the atom whether all matter was indeed composed of atoms and molecules. Einstein had suggested that by observing the erratic motion of pollen particles in a liquid, one could conclude that the warm liquid was indeed composed of molecules which were in constant motion at a finite temperature and that this motion produced random kicks on the pollen grain causing it to move along an erratic trajectory. Brownian motion had been observed several years ago by Robert Brown but it was generally believed that this motion was due to the organic matter having some sort of life, never was it believed that this motion was due to the kicks produced by the molecules at finite temperature and Einstein's theory was a bold step in this direction claiming Brownian motion to be a confirmation of the ultimate atomic structure of all matter. Today we know very well that Einstein was right. The model suggested here is as follows: Suppose a particle of mass M moving at time τ_k with a velocity $V(\tau_k-)$ suffered a collision the another particle of mass m moving with a velocity v_k. Then, just after the collision, the particle of mass M would move with a velocity $V(\tau_k+)$ where by conservation of momentum and energy,

$$MV(\tau_k-) + mv_k = MV(\tau_k+) + mv_k',$$

$$MV(\tau_k-)^2/2 + mv_k^2/2 = MV(\tau_k+)^2/2 + mv_k'^2/2$$

with v_k' being the velocity of the particle m just after the collision. This system of equations can then be solved to yield $V(\tau_k+)$ in as a linear combination of $V(\tau_k-)$ and v_k with the linear combination coefficients being some functions of M and m. Now we assume that the successive collision times $\tau_k, k = 1, 2, \ldots$ of particles of mass m moving with velocities $v_k, k = 1, 2, \ldots$ are the arrival times of a renewal process, ie, the inter-collision times $\tau_{k+1} - \tau_k, k = 0, 1, 2, \ldots$ are iid random variables with a distribution F and that the corresponding colliding velocities $v_k, k = 1, 2, \ldots$ are also iid random variables with a distribution G, then using the assumption that in between two successive collision times τ_k and τ_{k+1}, the particle M moved with a uniform velocity so that its velocity at time t was given by

$$V(t) = V(\tau_k+), \tau_k \geq t < \tau_{k+1}, V(\tau_{k+1}-) = V(\tau_k+)$$

we can in principle calculate the statistics of the velocity process $V(t)$ of M in terms of F and G and M and m. Then, by taking appropriate limits with appropriate choices of F and G (like F being the exponential distribution corresponding to Poisson collision times and G being the Maxwellian Gaussian velocity distribution), is it possible to arrive at Langevins' theory of the velocity process:

$$dV(t) = -\gamma V(t)dt + \sigma dB(t)$$

so that $V(t)$ is a Gaussian process with zero mean and autocorrelation $(\sigma^2/2\gamma)exp(-t_2|)$ when equilibrium has been reached ?

8.8   The Born-Oppenheimer program

Involves developing software packages for calculating the energy levels and stationary state wave functions for any kind of atom or molecule. Let me introduce this circle of ideas. The first well known approximate method for solving many atom problems was developed by Born and Oppenheimer. Consider a crystal having N nuclei, each of mass M and charge Ze. Assume that each nucleus has Z electrons, with the effective mass of each electron being m. Let P_n denote the momentum operator of the nth nucleus and let p_{ni} denote the momentum operator of the ith electron in the nth nucleus. Let R_n denote the position operator of the nth nucleus and let r_{ni} denote the position operator of the ith electron of the nth nucleus. Let $U_1(R_n, r_{mi})$ denote the interaction potential energy between the nth nucleus and the ith electron of the mth nucleus. Let $U_2(R_n, R_m)$ denote the interaction potential energy between the nth and the mth nucleus and finally, let $U_3(r_{ni}, r_{mj})$ denote the interaction potential energy between the ith electron of the nth nucleus and the jth electron of the mth nucleus. The total kinetic energy of the electrons is

$$T_e = \sum_{n,i} p_{ni}^2/2m$$

The total kinetic energy of the nuclei is

$$T_N = \sum_{n} P_n^2/2M$$

The total mutual interaction potential energy of the electrons is

$$U_e = \sum_{nmij} U_3(r_{ni}, r_{mj}) = U_e(\mathbf{r})$$

The total mutual interaction potential energy of the nuclei is

$$U_N = \sum_{n,m} U_2(R_n, R_m) = U_N(\mathbf{R})$$

and the total interaction potential energy between the nuclei and the electrons is

$$U_{Ne} = \sum_{nmi} U_1(R_n, r_{mi}) = U_{Ne}(\mathbf{R}, \mathbf{r})$$

The total Hamiltonian (energy operator) of the crystal is

$$H = T_N + T_e + U_N + U_e + U_{Ne}$$

Our aim is to solve the eigenvalue problem

$$H\psi(\mathbf{R}, \mathbf{r}) = E\psi(\mathbf{R}, \mathbf{r}) --- (1)$$

To this end, we assume a factorization of the wave function

$$\psi(\mathbf{R}, \mathbf{r}) = \phi(\mathbf{R})\chi(\mathbf{R}, \mathbf{r})$$

where $\chi(\mathbf{R}, \mathbf{r})$ is the wave function of the electrons alone with the nuclei at fixed positions:

$$(T_e + U_e + U_{Ne})\chi(\mathbf{R}, \mathbf{r}) = E_e(\mathbf{R})\chi(\mathbf{R}, \mathbf{r}) --(2)$$

$E_e(\mathbf{R})$ is the corresponding energy level of the electrons with the nuclei at the fixed positions $\mathbf{R}$. Substituting this into (1) and noting that T_e does not act on $\phi(\mathbf{R})$ gives us

$$(T_N + E_e(\mathbf{R}) + U_N(\mathbf{R}))\phi(\mathbf{R})\chi(\mathbf{R}, \mathbf{r}) = E\phi(\mathbf{R})\chi(\mathbf{R}, \mathbf{r})$$

To a first order of approximation, one then assumes that the action of the kinetic energy T_N of the nuclei on $\chi(\mathbf{R}, \mathbf{r})$ is negligible in view of the heaviness of the masses of the nuclei, ie, $M >> m$. Then one arrives at the approximate equation

$$(T_N + U_N(\mathbf{R}) + E_e(\mathbf{R}))\phi(\mathbf{R}) = E\phi(\mathbf{R}) --(3)$$

This equation tells us that $\phi(\mathbf{R})$ is the approximate wave function of the nuclei moving in the potential generated by their mutual interaction plus the energy of the electrons with the nuclei at fixed positions. If we did not make any such approximation, then

$$T_N\phi(\mathbf{R})\chi(\mathbf{R}, \mathbf{r}) =$$

$$T_N\phi(\mathbf{R}))\chi(\mathbf{R}, \mathbf{r}) + \phi(\mathbf{R})T_N\chi(\mathbf{R}, \mathbf{r}) +$$

$$\sum_n (P_n\phi(\mathbf{R})).(P_n\chi(\mathbf{R}, \mathbf{r})/M$$

and then the exact equation for $\phi(\mathbf{R})$ would read

$$(T_N + U_N(\mathbf{R}) + E_e(\mathbf{R}) - E)\phi(\mathbf{R})$$

$$+[\phi(\mathbf{R})(T_N\chi(\mathbf{R}, \mathbf{r}))/\chi(\mathbf{R}, \mathbf{r}) + (1/\chi(\mathbf{R}, \mathbf{r}))\sum_n (P_n\phi(\mathbf{R})).(P_n\chi(\mathbf{R}, \mathbf{r})/M] = 0 ---- (4)$$

If we solve the approximate nuclear eigenvalue problem for $\phi(\mathbf{R})$ and E as in (3), we can get the next order approximation by using (3) and (4) with E replaced by $E + \delta E$ and $\phi(\mathbf{R})$ replaced by $\phi(\mathbf{R}) + \delta\phi(\mathbf{R})$ in (4) and using first order perturbation theory to get

$$(T_N + U_N(\mathbf{R}) + E_e(\mathbf{R}) - E)\delta\phi(\mathbf{R}) - \delta E.\phi(\mathbf{R})$$

$$+[\phi(\mathbf{R})(T_N\chi(\mathbf{R},\mathbf{r}))/\chi(\mathbf{R},\mathbf{r}) + \sum_n (P_n\phi(\mathbf{R})).(P_n\chi(\mathbf{R},\mathbf{r})/M] = 0 - - - (5)$$

The dependence of this equation on $\mathbf{r}$ can be eliminated by multiplying both sides with $\chi(\mathbf{R},r)^*$ and integrating over $\mathbf{r}$:

$$(T_N + U_N(\mathbf{R}) + E_e(\mathbf{R}) - E)\delta\phi(\mathbf{R}) - \delta E.\phi(\mathbf{R})$$

$$+\phi(\mathbf{R})\int \chi(\mathbf{R},\mathbf{r})^* T_N \chi(\mathbf{R},\mathbf{r})d^3\mathbf{r}$$

$$+\sum_n (M^{-1}\int \chi(\mathbf{R},\mathbf{r})^* P_n\chi(\mathbf{R},\mathbf{r})d^3\mathbf{r})(P_n\phi(\mathbf{R})) = 0 - - - (6)$$

As in standard time independent first order perturbation theory in quantum mechanics, (6) can be solved for $\delta\phi(\mathbf{R})$ and δE using all the normalized nuclear eigenfunctions $\phi(\mathbf{R})$ corresponding to the different eigenvalues E in (3). The perturbation process can then be continued in (6) by replacing E and ϕ with $E + \delta E$ and $\phi + \delta\phi$.

The exact band structure of a semiconductor crystal could be calculated by a definite algorithm and this algorithm developed by Born and Oppenheimer several years ago .

8.9   The superposition principle for wave functions of the curved space-time metric field could lead to contradictions and what are the fundamental difficulties in developing a background independent theory of quantum gravity

8.10   Attempts to detect gravitational waves from rotating pulsars and sudden burst of a star using crystal detectors

All such attempts they mentioned had failed to detect gravitational waves and hence there was a big question mark on whether such waves predicted by Einstein's general theory of relativity really existed. From this time, it has now taken about twenty five years to conclusively prove the existence of gravitational waves thereby opening the way for designing more experiments to detect the graviton as the fundamental particle that transports gravitational waves just as the photon is the fundamental particle that transports electromagnetic waves and likewise the massive W and Z bosons are the fundamental particles that propagate the nuclear forces. Not surprisingly, the detection of gravitational waves came not via the effect of gravity on matter but rather its effect on electromagnetic waves coming from a laser. Einstein had shown that gravity can cause light to bend and can also affect the electromagnetic wave patterns in space-time, the former by virtue of the null geodesic equations in curved space-time and the latter by virtue of the covariant derivative appearing in the Maxwell equations in order that these equations be tensor equations, ie, valid for all observers in the universe.

8.11 Sketch of the proof of Shannon's coding theorems

Let A be an alphabet with a symbols and let p be a probability distribution on A. This defines the source. The channel is characterized by the transition probabilities $q_x(y), x \in A, y \in B$ where B is another alphabet with b symbols. The source is assumed to be memoryless which means that the letters outputted by it are iid random variables each one having the distribution p. This means that the probability of the source outputting the sequence $u = (u(1)u(2)..u(n)) \in A^n$ is given by $p_n(u) = \Pi_{i=1}^n p(u(i))$. The channel is also assumed to be memoryless which means that if the source emits the string $u = (u(1)...u(n))$, then the probability that the receiver receives the sequence $v = (v(1)...v(n)) \in B^n$ is

$$q_{u,n}(v) = \Pi_{i=1}^n q_{u(i)}(v(i))$$

From the Chebyshev inequality, we know that if $\delta > 0$ is an arbitrarily small positive real number, then the probability that the source will emit an n-long string u in the set

$$T(n,p,\delta) = \{u : |N(x|u) - np(x)| < \delta\sqrt{\delta n p(x)(1 - p(x))}\forall x \in A\}$$

can be made arbitrarily close to unity by choosing n large enough and further, the probability of each sequence in $T(n,p,\delta)$ is arbitarily close to $2^{-nH(p)}$ for large n (on a logarithmic scale) and the number of sequences in $T(n,p,\delta)$ is likewise arbitrarily close to $2^{nH(p)}$. In other words for large n, a sequence emitted by the source will almost completely surely fall in $T(n,p,\delta)$ called the set of δ-typical sequences and further that these sequences will be nearly uniformly distributed over $T(n,p,\delta)$ with each sequence having probability $2^{-nH(p)}$. This result form the core of Shannon's noiseless coding theorem that is the basis of data compression. Likewise, since the probability that the received symbol is y is given by $q(y) = \sum_{x \in A} p(x)q_x(y)$, it follows that the received sequence in B^n for sufficiently large n will almost certainly fall in $T(n,q,\epsilon)$ with ϵ arbitrarily small when n large enough. The number of n-long typical sequences transmitted is $2^{-nH(p)}$ while the number of n-long typical sequences received is $2^{-nH(q)}$. Further, if $u(i)$ is the ith symbol transmitted and $v(i)$ the corresponding symbol received, then the pairs $(u(i), v(i)), i = 1, 2, ...$ are iid with probability distribution $\mu(x, y) = p(x)q_x(y)$ and we can also therefore talk about jointly typical sequences in transmitted-received symbol pair space. The transmitted-received sequence pair will also be jointly typical with probability nearly one. Moreover the typical sequences are nearly uniformly distributed over the typical sequence space and hence given a transmitted sequence and a received sequence in their respective typical spaces, the probability of the corresponding joint sequence being typical is

$$\frac{2^{nH(X,Y)}}{2^{nH(X)}2^{nH(Y)}} = 2^{-nI(X,Y)}$$

where

$$H(X, Y) = H(X) + H(Y) - H(X, Y)$$

is the mutual information between a transmitted and received symbol. Here X is a random variable with values in A having distribution p so that $H(X) = H(p)$, Y is a random variable in B having distribution q so that $H(Y) = H(q)$ and $H(X, Y) = H(\mu)$. We encode the n-long strings transmitted by the source in the following way. Let $R < H(X)$. Then one of 2^{nR} distinct messages $\{x_1, ..., x_{2^{nR}}\}$ selected at random from the $2^{nH(X)}$ typical sequences of the source is transmitted. Then when the ith message x_i is transmitted and y_i is the corresponding received message $(i = 1, 2, ..., 2^{nR})$, the receiver decides that that x_j has been transmitted if (x_j, y_i) is jointly typical. A decoding error will therefore occur if either (x_i, y_i) is not jointly typical or else if (x_j, y_i) is typical for some $j \neq i$. The average error probability with this random code is then given by

$$2^{-nR} \sum_{i=1}^{2^{nR}} P((x_i, y_i) is not typical \, or \, (x_j, y_i) is typical for some j \neq i | y_i)$$

where the received sequence y_i is typical. The probability of (x_i, y_i) being non typical given a typical output sequence y_i is negligible while the probability of (x_j, y_i) being typical given that both x_j, y_i are typical is $2^{-nI(X,Y)}$ as we saw above. Thus, the average error probability of this random code is smaller than (using the union bound)

$$2^{n(R-I(X,Y))}$$

and this probability will converge to zero as $n \to \infty$ provided that $R < I(X, Y)$.

8.12 The notion of a field operator or rather an operator valued field

How this idea could be used to develop the matrix mechanics of Heisenberg for the electromagnetic field of photons interacting with the electron-positron field of Dirac. This process of making classical fields like the electromagnetic four potential and the Dirac wave function into field operators and then introducing canonical commutation relations (CCR) for Bosonic fields and canonical anticommutation relations (CAR) for Fermionic fields is a new idea. For example, the CCR for a system of particles reads $[q_a(t), p_b(t)] = i\delta_{ab}$ while the CCR for position and momentum fields in space-time reads $[Q_a(t,x), P_b(t,y)] = i\delta_{ab}\delta^3(x-y)$ which means that rigorously speaking, the field operators are actually operator valued distributions. To give meaning to this CCR for fields, one uses test functions $\phi(x), \psi(x)$ which are rapidly decreasing functions and replaces the position and momentum field respectively by

$$Q_a(\phi) = \int Q_a(t,x)\phi(x)d^3x, P_a(\phi) = \int P_b(t,x)\phi(x)d^3x$$

and then the CCR assumes the form

$$[Q_a(\phi), P_b(\psi)] = i\delta_{ab} < \phi, \psi >$$

where

$$< \phi, \psi >= \int_{\mathbb{R}^3} \phi(x)\psi(x)d^3x$$

This means that by using test functions, the position and momentum fields are smoothened out thereby facilitating the definition of the CCR. This idea had been presented in my father's book in the formalism of creation and annihilation fields derived from position and momentum fields. Specifically by considering an infinite sequence of independent quantum harmonic oscillators with creation operators a_n^* and annihilation operators $a_n, n = 1, 2, \ldots$ obeying the Bosonic CCR

$$[a_n, a_m^*] = \delta_{nm}$$

we can construct annihilation and creation fields

$$a(\phi) = \sum_n a_n < \phi, e_n >, a(\phi)^* = \sum_n a_n^* < e_n, \phi >$$

where $\phi \in L^2(\mathbb{R}_+)$ and $e_n, n = 1, 2, \ldots$ is an orthonormal basis for $L^2(\mathbb{R}_+)$. The CCR's can now be expressed as

$$[a(\phi), a(\psi)^*] = \sum_n < \phi, e_n >< e_n, \psi >=< \phi, \psi >= \int_0^\infty \bar{\phi}(t)\psi(t)dt$$

In the quantum noise theory of Hudson and Parthasarathy (my father), the crucial step in creating a quantum noise process was to take the functions ϕ and ψ as the indicator function $\chi_{[0,t]}$ and define operator valued processes

$$A(t) = a(\chi_{[0,t]}), A(t)^* = a(\chi_{[0,t]})^*$$

and then prove the celebrated quantum Ito formula

$$dA(t).dA(t)^* = dt, dA(t)^*dA(t) = (dA(t))^2 = (dA(t)^*)^2 = 0$$

The proof of these formulae hinges around the CCR. The CCR's for the creation and annihilation operators are in turn consequences of the CCR's for position and momentum operators of a sequence of independent harmonic oscillators and hence they reflect the Heisenberg uncertainty principle. That the Ito's formula for classical Brownian motion is a special case of this quantum Ito formula once one recognizes that the operator valued process $B(t) = A(t) + A(t)^*, t \geq 0$ is commutative and its quantum statistics in a vacuum coherent state is the same as that of classical standard Brownian motion means that the Ito formula for Brownian motion can be regarded as a manifestation of the Heisenberg uncertainty principle !

In the first volume of Wienberg's book, the electromagnetic potentials are expanded in terms of creation and annihilation fields in 3-momentum space, ie, we have formulas like

$$A^m(t, \mathbf{r}) = \int (2|\mathbf{K}|)^{-1/2}(|a(\mathbf{K})e^m(\mathbf{K})exp(-i(|K|t - \mathbf{K}.\mathbf{r})) +$$

$$a(\mathbf{K})^* e^m(\mathbf{K})^* exp(i(|\mathbf{K}|t - \mathbf{K}.\mathbf{r})))d^3 K$$

Such quantum fields do not satisfy the Quantum Ito formula and hence cannot be used to model quantum noise. However, if we replace in this factor the sinusoidal modulating functions $exp(\pm i|\mathbf{K}|t)$ by indicator functions $\chi_{[0,t]}$ inner producted with sinusoids, ie,

$$\int_0^t exp(\pm i|\mathbf{K}|s)ds$$

as a substitute for the term $< \chi_{[0,t]}, e_n >$, then we can use the entire edifice of quantum field theory to describe quantum noise. It took me a very long time to realize all this. Wienberg's first volume on the quantum theory of fields, gave me the required stimulus to understand the computation of amplitudes of scattering, absorption and emission processes for elementary particles using the Feynman diagrammatic technique. How such amplitudes are computed and how these computations are related to the S-matrix of operator theory using the wave operators and S-matrix defined by a pair of Hamiltonians, the first Hamiltonian corresponding to that of a free projectile and the second Hamiltonian corresponding to that of the free projectile plus its interaction potential energy with the scattering centre is the crucial part here. In order to a jump from this idea of wave operator theory based on two Hamiltonians to quantum field theory, one must replace the free projectile Hamiltonian by the Hamiltonian of the free electromagnetic field plus that of the free Dirac field of electrons and positrons in the second quantized picture and the interaction Hamiltonian by that of the interaction Hamiltonian between the Dirac current field and the electromagnetic field. Then, one must write down the Dyson series for the unitary evolution operator corresponding to the sum of these two Hamiltonians in the interaction picture and then identify from each term in this Dyson series, a set of Feynman diagrams that describe the amplitudes of a given scattering process. In short, it is the Dyson series or equivalently higher order time independent perturbation theory that provides the link between the operator theoretic description of the scattering matrix as described in books on functional analysis and the practical tool of Feynman diagrams used by physicists to compute scattering amplitudes. A further development in this direction is renormalization theory, a method by introducing scaling of particle masses, charges and fields in terms of ultraviolet and infrared cutoffs introduced in order to obtain finite values of scattering amplitudes. Till today, it is a mystery that how by avoiding infinities in integrals used to evaluate amplitudes using renormalization methods, we are able to obtain answers that tally with experiments in particle accelerators. One of the most striking developments in this direction is quantum gravity wherein the metric field is also quantized and then the action function of all the other particle fields and the gravitational field interacting with each other is set up and the Feynman diagrammatic rules are applied to this problem after appropriate truncation of the components involving the metric field. Einsteinian gravity is not a renormalizable theory because its action involves all powers of the metric perturbations not just upto four as is required for a renormaliable theory. The electromagnetic and Dirac fields are renormalizable theories and since Einsteinian gravity is not, we have to artificially impose truncations, a method that destroys the diffeomorphism invariance of the theory. Even Feynman had written some notes on the application of the diagrammatic method to gravity interacting with the other particles after appropriate truncation but was not satisfied with it. The first major attempt at quantizing gravity which has scored mathematical success but not experimental success is superstring theory. Whereas gravity is an impossibility in conventional quantum field theory owing to renormalization problems, it becomes in quantum string theory an inevitable consequence by virtue of the string action being forced to obey conformal invariance. If one considers the Bosonic string action on a two dimensional world sheet taking into account the metric tensor of 26-dimensional space-time to obtain a diffeomorphic invariant string action, then on quantizing the string field by introducing a string propagator, one can deduce that for conformal invariance of the quantum average of the string action, where by conformal invariance, we mean invariance under multiplication of the string world-sheet metric by an arbitary positive function of the two world-sheet coordinates of the string, we end up with Einstein's field equation for gravitation $R_{ij} = 0$ in 26-dimensional space-time. This is indeed remarkable since its says that gravity appears as an inevitable consequence of quantization of the string when conformal invariance of the action is imposed. This is perhaps one of the most remarkable results in mathematical physics.

8.13   Group theoretic Pattern recognition

The problem is that we are given N image pattern fields, say $f_1, ..., f_N$. These pattern fields are defined as functions on a curved manifold on which a group G of transformations acts. Now we transform the first pattern f_1 by some element g_1 of the group, the second by g_2 and so on the Nth by g_N. The resulting transformed patterns are $g_1.f_1, g_2.f_2, ..., g_N.f_N$ where $g \in G$ acts on a pattern f as a group representation. For example, if a pattern f is represented as a function $f(x), x \in \mathcal{M}$ on the manifold $\mathcal{M}$, then its transformed by $g \in G$ to the function $(g.f)(x) = f(g^{-1}x), x \in \mathcal{M}$. This map on the space of patterns to itself defines a representation U of G. Thus, we write $g.f(x) = (U(g)f)(x) = f(g^{-1}x)$. We can add patterns and multiply a pattern by a real or complex scalar to get a new pattern. Thus, the space of patterns is closed under finite linear combinations. In other words, the space of patterns on the manifold $\mathcal{M}$ is a vector or linear space. We introduce a G-invariant measure on the manifold $\mathcal{M}$ that is induced by the left invariant Haar measure on G. Denoting this measure by μ, we consider only those patterns f on $\mathcal{M}$ that have finite energy w.r.t the measure μ, ie, those functions $f(x)$ on $\mathcal{M}$ that are Borel measurable and for which $\int_{\mathcal{M}} |f(x)|^2 d\mu(x) < \infty$. This condition can equivalently be stated as $\| f \|^2 = \int_G |f(g.x_0)|^2 dg < \infty$ where $x_0 \in \mathcal{M}$ is some fixed point in $\mathcal{M}$ and dg is the left invariant Haar measure on $\mathcal{M}$. This condition will not depend on x_0 provided that G acts transitively on $\mathcal{M}$, ie, for any two points $x, y \in \mathcal{M}$, there is a $g \in G$ for which $g.x = y$. With the norm $\| f \|$ of a pattern f defined in this way, the space of finite norm patterns on $\mathcal{M}$ becomes a Hilbert space $\mathcal{H} = L^2(\mathcal{M}, \mu)$, ie, a linear space with a scalar product: $< f_1, f_2 > = \int_{\mathcal{M}} \bar{f_1}(x) f_2(x) dx = \int_G \bar{f_1}(gx_0) f_2(gx_0) dg$ and under this norm, $\mathcal{H}$ is complete, ie, every Cauchy sequence converges. The action U of G in $\mathcal{H}$ then becomes a unitary representation, ie,

$$\| U(g)f \| = \| f \|, g \in G, f \in \mathcal{M}$$

ie $U(g)$ preserves the norm of any finite norm pattern or equivalently, the scalar product between any two finite norm patterns. Now, the question is that from the G-transformed patterns, we wish to recover the original pattern, or more precisely, we wish to identify which of the original patterns $f_1, ..., f_N$ was transformed to give the given measured f ? This question can be answered if we are able to define a set of invariants $I_l : \mathcal{H} \to \mathbb{C}, l \in J$ for the representation U, ie, each $I_l, l \in J$ should map every pattern $f \in \mathcal{H} = L^2(\mathcal{M})$ into a real or complex number such that $I_l(U(g)f) = I_l(f) \forall g \in G, f \in \mathcal{H}$ in such a way that given any two distinct $f_1, f_2 \in \mathcal{H}$, ie, such that $\| f_1 - f_2 \| > 0$ and such that there does not exist any $g \in G$ for which $f_2 = U(g)f_1$, ie, $\| f_2 - U(g)f_1 \| > 0 \forall g \in G$, there should be at least one $l \in J$ such that $I_l(f_1) \neq I_l(f_2)$. This is equivalent to saying that each I_l should be a constant on each G-orbit of $\mathcal{H}$, ie the $I_l's$ should be G-invariants and that this class of invariants $I_l, l \in J$ should separate $\mathcal{H}$. I observed during my work on this project that for the three dimensional rotation group acting on the sphere, we can easily construct such invariants by taking the norm square of the projection of a pattern onto the different irreducible subspaces. More precisely, let S^2 denote the surface of the three dimensional unit sphere. $SO(3) = G$ is the three dimensional rotation group that acts on the Hilbert space $L^2(S^2)$ of finite energy patterns defined on S^2. The Haar measure on $SO(3)$ induces the invariant area measure $sin(\theta)d\theta.d\phi$ on S^2. Let $Y_{lm}(\theta, \phi), m = -l, -l+1, ..., l-1, l, l = 0, 1, 2, ...$ denote the spherical harmonics. Define the finite dimensional Hilbert spaces $\mathcal{H}_l = span\{Y_{lm} : |m| \leq l\}$. let $U(g)f(x) = f(g^{-1}x), f \in L^2(S^2), g \in G = SO(3)x = (\theta, \phi) \in S^2$. Then it is well known that the operators $U(g), g \in G$ leave each $\mathcal{H}_l$ invariant, that the restrictions U_l of U to $\mathcal{H}_l$ are irreducible unitary representations of G and that the $U_l, l = 0, 1, 2, ...$ exhaust all the finite dimensional irreducible representations of G upto equivalence. Then, for each $f \in L^2(S^2) = \mathcal{H}$, define

$$P_l f(\theta, \phi) = \sum_{|m| \leq l} Y_{lm}(\theta, \phi) < Y_{lm}, f >$$

where

$$< u, v >= \int_0^{2\pi} \int_0^{\pi} \bar{u}(\theta, \phi)v(\theta, \phi)sin(\theta)d\theta.d\phi$$

Then P_l is the orthogonal projection of $\mathcal{H}$ onto $\mathcal{H}_l$ and

$$I_l(f) =\| P_l f \|^2= \sum_{|m| \leq l} | < Y_{lm}, f > |^2, l = 0, 1, 2, ...$$

form a set of G-invariants, ie,

$$I_l(U(g)f) = I_l(f), g \in G, f \in \mathcal{H}$$

Suppose that f_1 and f_2 fall in different G-orbits, then is it true that for some $l \geq 0$, we must necessarily have $I_l(f_2) \neq I_l(f_1)$. This problem has worried me for quite some time. The following facts emerge: Suppose $I_l(f_1) = I_l(f_2)$ for every $l \geq 0$. Then for each l, there exists a unitary operator $V_l : \mathcal{H}_l \to \mathcal{H}_l$ such that

$$P_l f_2 = V_l P_l f_1$$

Hence

$$f_2 = \sum_l P_l f_2 = U_0 f_1$$

where $U_0 : \mathcal{H} \to \mathcal{H}$ is the unitary operator

$$U_0 = \sum_l V_l P_l = \sum_l P_l V_l P_l$$

Since U_l is irreducible in $\mathcal{H}_l$, it follows that there is a complex valued function $a_l(g)$ on G such that

$$V_l = \int_G a_l(g)U_l(g)dg$$

It is not clear from these considerations whether the set $I_l, l \geq 0$ of invariants is complete or not, ie, whether there exists a $g_0 \in G$ such that $V_l = U_l(g_0)\forall l$. If there does exist such a g_0, then it would follow that $U_0 = U(g_0)$ and hence the completeness of the $I_l's$.

8.14 Controlling the probability distribution in functional space of the Klein-Gordon field using a field dependent potential

Let $f(x, \phi(x))$ be the potential. Here, $x = (t, r)$ is a space-time coordinate and $\phi(x)$ is the KG field after it has got perturbed by the potential f. More generally, in super-symmetry theory, f is a superpotential that must be gauge invariant, ie, if t_A are the generators of the gauge group, then

$$\sum_n \frac{\partial f(\phi)}{\partial \phi_n}(t_A\phi)_n = 0 \forall \phi, A$$

or equivalently in terms of matrix elements,

$$\sum_{n,m} \frac{\partial f(\phi)}{\partial \phi_n}(t_A)_{nm}\phi_m = 0 \forall \phi, A$$

These constraints on the super-potential can be realized by fixing its function form except for a finite set of unknown parameters which may be fine tuned with time so that the resultant probability density functional of the components of the superfield tracks a given probability density functional.

When the Klein-Gordon Lagrangian is perturbed by a potential, the field equations become

$$\nabla^2\phi - \phi_{,tt} - \mu^2\phi - f(x, \phi) = 0$$

We write

$$f(x, \phi) = \sum_n f_n(t|\theta)\phi^n$$

where now the functions $f(x|\theta) = f(t, r|\theta)$ are known except for the control parameters θ. We consider this field confined to the volume within a cube of length L. Expanding the field $\phi(t, r)$ in spatial Fourier series within this cube gives us

$$\phi(t, r) = \sum_n c_n(t)exp(2\pi in.r/L)$$

where $n = (n_1, n_2, n_3) \in \mathbb{Z}^3$. Substituting this into the above KG equation gives us

$$(2\pi/L)^2 n^2 c_n(t) + c_n''(t) + \mu^2 c_n(t) - \sum_{m_1,...,m_r,r:m_1+...+m_r=n} f_{m_1}(t|\theta)...f_{m_r}(t|\theta)c_{m_1}(t)...c_{m_r}(t) = 0$$

Here, we are assuming that f_n depends on time but not on the spatial coordinates.

8.15 Quantum processing of classical image fields using a classical neural network

Given an $N \times N$ image field of pixels, with the $(i,j)^{th}$ pixel having an intensity $I(i,j) \in [0,1]$, we encode the intensity of the pixel into a single qubit state

$$|i,j> = I(i,j)exp(i\phi(i,j))|1> + \sqrt{1 - I(i,j)^2}exp(i\psi(i,j))|0>$$

where $\phi(i,j)$ and $\psi(i,j)$ are arbitary phase factors introduced to increase the number of processing degrees of freedom of the image. The qubit $|1>$ stands for the maximum intensity, namely white and the qubit $|0>$ for the minimum intensity, namely black. Given that the $(i,j)^{th}$ pixel is in the state $|i,j>$ after this classical to quantum encoding process, the probability that a measurement will yield the brightest state $1>$ is $|<1|i,j>|^2 = I(i,j)$ and the probability that the measurement will yield the darkest state $|0>$ is $|<0|i,j>|^2 = 1 - I(i,j)$. Thus, if the pixel is classically bright, there is more probability of it being in the state $|1>$ and if it is classically dark, there is more probability of it being in the state $|0>$.

8.16 Entropy and supersymmetry

Consider a supersymmetric Lagrangian $L(\phi, \phi_{,\mu})$ where ϕ is a set of component superfields. For example, we can take

$$L = [K(\Phi^*, \Phi)]_D$$

where Φ is a left Chiraal superfield with component superfields ϕ, for example, if $\Phi_n, n = 1, 2, ..., N$ are N left Chiral superfields, we can take

$$L = \sum_n [\Phi_n^* \Phi_n]_D$$

This Lagrangian will contain the kinetic energy terms of the Klein-Gordon scalar field, kinetic energy terms of the Dirac field and some auxiliary field terms. To get the potential energy terms and gauge field interaction terms in the scalar and Dirac field, we must introduce a gauge superfield V and replace L by

$$L = [\Phi^*.exp(V)\Phi]_D + [f(\Phi)]_F + [Tr(W_L^T \epsilon W_L)]_F$$

where f is a real valued function called the superpotential. We can also make f to depend explicitly on the space-time coordinates $x = (t, r)$ and thereby control the superpotential. W_L is a left Chiral superfield constructed using the left and right superderivatives of the gauge superfield V. The last term above is a gauge invariant action and also of course supersymmetric since it is the F component of a left Chiral superfield. The gauge invariance of this term is a consequence of the transformation of the generalized gauge transformation of the gauge super field $exp(V)$:

$$exp(V) \rightarrow exp(i\Omega(x_+, \theta_L))exp(V)exp(-i\Omega(x_+, \theta_L)^*)$$

(Note that V is not a Chiral superfield) and hence

$$exp(-V) \rightarrow exp(i\Omega(x_+, \theta_L)^*)exp(-V).exp(-i\Omega(x_+, \theta_L))$$

$\Omega(x_+, \theta_L)$ represents the most general left Chiral superfield. Using the fact that the right superderivatives D_R annihilate the left Chiral supefield $\Omega(x_+, \theta_L)$ while the left superderivatives D_L annihilate the right Chiral superfield $\Omega(x_+, \theta_L)^*$, it can be readily inferred from the construction of W_L in terms of V that under a generalized gauge transformation, W_L transforms as

$$W_L \rightarrow exp(i\Omega(x_+, \theta_L))W_L.exp(-i\Omega(x_+, \theta_L))$$

and hence that $[Tr(W_L^T \epsilon W_L)]_F$ is gauge invariant. Note that the gauge superfield V is of the form $V_A(x, \theta)t_A$ (summation over the Yang-Mills index A) where t_A are the Hermitian generators of the Yang-Mills gauge group. The first term $[\Phi^*.exp(V)\Phi]_D$ contains the kinetic energy terms of the matter fields, namely the scalar KG field and the Dirac field and also the interaction terms of these fields with the gauge fields while the term $[Tr(W_L^T \epsilon W_L)]_F$ contains the Lagrangian of the gauge fields and their superpartners, the gaugino fields and the auxiliary fields. The superpotential $[f(\Phi)]_F$ contains the potential energy of the matter fields which give masses to the matter fields. By controlling the superpotential, we effectively control the masses of the matter field particles. It is just like controlling the environment in which these masses move thereby causing the effective masses of these particles to get shifted.

The Lagrangian will then have the form

$$L = \sum_{k \geq 0} g_k(t, r) L_k(\phi, \phi_{,\mu}, V_\mu, V_{\mu,\nu})$$

where $g_0(t, r) = 1$ and L_0 consists of only the matter and gauge field Lagrangians without their interactions. More generally, we can after passing over to Hamiltonians using the Legendre transformations, consider Hamiltonians of the form

$$H = H_0 + V_0 + \sum_{k \geq 1} g_k(t) V_k$$

where H_0 is the unperturbed Hamiltonian consisting of only the Hamiltonian of the Klein -Gordon field, the Dirac field and the electromagnetic field. V_0 is a perturbation Hamiltonians consisting of the interactions between the Klein-Gordon field and the electromagnetic field and between the Dirac field and the electromagnetic field. $\sum_{k \geq 1} g_k(t) V_k$ consists of the interaction Hamiltonian between the Klein-Gordon field and a classical control electromagnetic potential, interaction between the Dirac field and the classical control electromagnetic potential and interaction between the electromagnetic field and a classical control current source field.

These terms have the form

$$H_0 = \int [(1/2)(\partial_t \phi)^2 + (1/2)(\nabla \phi)^2 + m^2 \phi^2 / 2$$

$$+\psi^*((\alpha, -i\nabla) + \beta m)\psi - (1/4)F_{\mu\nu}F^{\mu\nu})d^3r$$

and

$$V_0 = -e\psi^* \alpha^\mu \psi . A_\mu + (1/2)[(\partial^\mu + ieA^\mu)\phi.(\partial_\mu - ieA_\mu)\phi]_{int}$$

where $[X]_{int}$ stands only for the interaction part, ie,

$$[(\partial^\mu + ieA^\mu)\phi.(\partial_\mu - ieA_\mu)\phi]_{int} =$$

$$[(\partial^\mu + ieA^\mu)\phi.(\partial_\mu - ieA_\mu)\phi] - [\partial^\mu \phi . \partial_\mu \phi]$$

$$= e^2 A_\mu A^\mu \phi^2 - 2eIm(A^\mu \partial_\mu \phi)$$

More generally in the gauge field terms, we can also include non-Abelian gauge fields. Now we come to the question of how much entropy does the gauge field interaction plus the external c-number stochastic field interaction pump into the matter field. To answer this question, we must express the free gauge field in terms of Boson creation and annihilation operators and also express the free matter fields, namely the KG and Dirac fields respectively in terms of Boson creation and annihilation operators and Fermion creation and annihilation operators and thereby express the interaction components in the Hamiltonian in terms of these Boson and Fermion creation and annihilation operators and their couplings with the classical stochastic c-number control fields. We then compute the change in the mixed state of the system under such an interaction and then calculate the the Von-Neumann entropy of this changed state. Sometimes under such interactions, a pure state can transform into a mixed state. This happens for example in quantum blackhole physics. In quantum blackhole physics, the particles moving in the vicinity of the blackhole like the photons, electrons, positrons, neutrinos, the nuclear particles, the gauge bosons that propagate the nuclear forces etc., are initially in pure states. The ensemble of all these particles is described by their joint Hamiltonian. However, the gravitons which are generated by the blackhole are bosons with a Hilbert space described by a Boson Fock space and the associated graviton creation and annihilation operators. The gravitons, according to Einstein's general theory of relativity are spin two massless particles. The gravitons represent the bath to which the system comprising of the photons, electrons, positrons, neutrinos, the nuclear particles and the gauge bosons move. The initial state of the system is a pure state $|f >$ and the initial state of the bath is again a pure state, say a coherent state $|\phi(u) >$ in the graviton Fock space. The initial state of the system and bath is therefore the pure state $|f \otimes \phi(u) >$.

Chapter 9

Problems in Information Theory

[1] Given two discrete memory channels $\mathcal{C}_1$ and $\mathcal{C}_2$ characterized by the transition probability distributions $\nu_{x_1}^{(1)}(y_1)$ and $\nu_{x_2}^{(2)}(y_2)$ with $x_1 \in A_1, y_1 \in B_1, x_2 \in A_2, y_2 \in B_2$, then the product $\mathcal{C}_1 \times \mathcal{C}_2$ of these two channels is a discrete memoryless channel characterized by the transition probability distribution $\nu_{(x_1,x_2)}(y_1,y_2) = \nu_{x_1}(y_1)\nu_{x_2}(y_2)$ with input alphabet $A_1 \times A_2$ and output alphabet $B_1 \times B_2$. Let C_1 denote the capacity of $\mathcal{C}_1$ and C_2 that of $\mathcal{C}_2$. Theh prove that the capacity of $\mathcal{C}_1 \times \mathcal{C}_2$ is given by $C_1 + C_2$.

hint: Use the fact that if μ is a probability distribution on $B_1 \times B_2$ with marginals μ_1 and μ_2, then

$$H(\mu) \leq H(\mu_1) + H(\mu_2)$$

[2] Prove that if X, Y, Z are random variables on a fixed probability space, then

$$H(X|Y,Z) \leq H(X|Y)$$

Use this to deduce that if $\{X_n : n \in \mathbb{Z}\}$ is a stationary stochastic process, then the entropy rate

$$\bar{H} = lim_{n \to \infty} \frac{H(X_0, X_1, ..., X_n)}{n}$$

exists and equals $H(X_0|X_{-1}, X_{-2}, ...)$.

[3] Prove that $x \to log(x)$ is a concave function on $\mathbb{R}_+$ and that $x \to -x.log(x)$ is also a concave function on $\mathbb{R}_+$. Deduce from this the concavity of the entropy, ie, if $p = (p(1), ..., p(N))$ and $q = (q(1), ..., q(N))$ are two probability distributions on the set $A = \{1, 2, ..., N\}$ and $t \in [0, 1]$, then

$$H(tp + (1 - t)q) \geq tH(p) + (1 - t)H(q)$$

109

where $tp + (1-t)q$ is the probability distribution $\{tp(1) + (1-t)q(1), ..., tp(N) + (1-t)q(N)\}$ on $\{1, 2, ..., N\}$.

[4] Let μ be a probability distribution on $A \times B$ where A and B are two finite alphabets. Let μ_1 be the first marginal of μ and μ_2 the second marginal. Prove that

$$H(\mu) \leq H(\mu_1) + H(\mu_2)$$

using the fact that if p and q are two probability distributions on a finite set E, then

$$\sum_{x \in E} p(x)log(\frac{p(x)}{q(x)}) \geq 0$$

Deduce from this result that if (X, Y) is a pair of random variables having a joint distribution on a given probability space and assuming values in a finite set, then

$$I(X, Y) = H(X) + H(Y) - H(X, Y) \geq 0$$

with equality iff X and Y are independent random variables.

[5] if $X_n, n \in \mathbb{Z}$ is an ergodic stochastic process with probability distribution μ on the sequence space, then prove the Shannon-Mcmillan-Breiman theorem:

$$lim_{n \to \infty} \frac{-log(\mu(X_1, ... X_n))}{n} = \bar{H}(\mu) = H(X_0 | X_{-1}, X_{-2}, ...)$$

almost surely. Deduce from this result, Shannon's noiseless coding/data compression theorem for ergodic sources: Given $\epsilon, \delta > 0$, there exists a sufficiently large finite integer $N(\epsilon, \delta)$ such that for every $n > N(\epsilon, \delta)$, there is an $E_n \subset A^n$ such that

$$N(E_n) \leq 2^{n(\bar{H}(\mu) + \delta)}$$

and

$$\mu(E_n) > 1 - \epsilon$$

Further, deduce the converse, namely, if $E_n, n = 1, 2, ...$ is a sequence with $E_n \subset A^n$ such that $\mu(E_n) \to 1$, then

$$liminf_{n \to \infty} \frac{log(N(E_n))}{n} \geq \bar{H}(\mu)$$

In words, this means that the entropy rate is the best possible reliable compression of the data in terms of number of compressed bits per data bit.

[6] Construct the optimal Huffman code for a set of five source symbols with probabilities $p(j), j = 1, 2, ..., 5$ such that

$$p(1) \geq p(2) \geq p(3) \geq p(4) \geq p(5),$$

$$p(1) \geq p(2) \geq p(4) + p(5) \geq p(3),$$

$$p(3) + p(4) + p(5) \geq p(1) \geq p(2),$$
$$p(1) + p(2) \geq p(3) + p(4) + p(5)$$

Give an example of a probability distribution in which these inequalities are satisfied.

[7] Let A and B be two finite alphabets with $N(b) = b$ and let $S(B) = \bigcup_{n \geq 1} B^n$, denote the infinite set of all strings in B. A code for A with code alphabet B is given by a map $f : A \rightarrow S(B)$, ie, each $x \in A$ is mapped to a string $f(x)$ of B alphabets. The extension $\tilde{f}$ of a code f is the map $\tilde{f} : S(A) \rightarrow S(B)$ satisfying

$$f(x_1 x_2 ... x_n) = f(x_1) f(x_2) ... f(x_n), x_i \in A, i = 1, 2, ..., n, n = 1, 2, ...$$

f is said to be a uniquely decipherable code if given any two strings $x = (x_1 ... x_n)$ and $y = (y_1 ... y_m)$ in A^n and A^m respectively such that $m \geq n$, if x is not a prefix of y, then $\tilde{f}(x)$ must not be a prefix of $\tilde{f}(y)$. Show that a code f is uniquely decipherable iff $\sum_{x \in A} b^{-l(f(x))} \leq 1$ where $l(f(x))$ is the length of $f(x) \in S(B)$.

[8] Carry out the following steps in the proof of the ergodic theorem: Let $(\Omega, \mathcal{F}, P)$ be a probability space and $T : \Omega \rightarrow \Omega$ a measurable measure preserving transformation, ie, $T^{-1}(E) \in \mathcal{F} \forall E \in \mathcal{F}$ and $PT^{-1} = P$ on $\mathcal{F}$, the Birkhoff's individual ergodic theorem states that if $f \in L^1(\Omega, \mathcal{F}, P)$, ie, $\mathbb{E}(|f|) \int |f| dP < \infty$ then

$$lim_{n \rightarrow \infty} n^{-1} \sum_{i=0}^{n-1} f(T^i \omega) = f^*(\omega)$$

exists for P almost every $\omega \in \Omega$ and that $f^*(T\omega) = f^*(\omega)$ for P a.e ω, ie, f^* is an invariant function. Show that $f^* \in L^1$. Define the invariant σ-field

$$I = \{E \in \mathcal{F} : T^{-1}(E) = E\}$$

Then, show that f^* is I-measurable, ie, $f^{*-1}(B) \in I$ for all Borel subsets B of $\mathbb{R}$ in the almost sure sense, ie, $E = f^{*-1}(B)$, then

$$P(T^{-1}(E) \Delta E) = 0$$

Hence deduce that

$$f^*(\omega) = \mathbb{E}(f|I)(\omega)$$

by showing that if $E \in I$, then

$$\int_E f^* dP = \int_E f dP$$

For doing this part, you must first show that for $E \in I$,

$$\int_E f o T^i dP = \int_E f dP, i = 1, 2, ...$$

by using the facts $T^{-1}(E) = E, PoT^{-1} = P$ and the change of variable formula in integration. Now suppose, that in addition to being measure preserving, T is also ergodic, ie $I = \{\phi, \Omega\}$ in the almost sure sense, or more precisely, $E \in I$ implies $P(E) = 0$ or 1. Then deduce that

$$f^*(\omega) = \mathbb{E}(f) = \int f dP$$

ie f^* is a constant. To prove the ergodic theorem, we first define the partial sums

$$S_n = \sum_{i=0}^{n-1} foT^i, n \geq 1, S_0 = 0$$

and then

$$M_n = max(S_k : 0 \leq k \leq n)$$

Then one easily proves that on the set $\{M_n > 0\}$, one has

$$M_n = f + M_{n-1}oT \leq f + M_noT$$

and hence, since $M_n \geq 0$,

$$\int M_n dP = \int_{M_n>0} M_n dP \leq \int_{M_n>0} f dP + \int_{M_n>0} M_noT dP$$

$$\leq \int_{M_n>0} dP + \int M_n dP$$

thereby yielding the maximal ergodic theorem:

$$\int_{M_n>0} f dP \geq 0$$

Replacing f by $f\chi_E$ where E is any invariant set in this argument, one deduces that

$$\int_{M_n>0 \cap E} f dP \geq 0$$

Now consider for $a > 0$ the set

$$E_a = \{sup_{n\geq 1} n^{-1} S_n > a\} = \{sup_{n\geq 1} \tilde{S}_n > 0\}$$

where $\tilde{S}_n$ is obtained by using $g = f - a$ in place of f in the definition of S_n. Then, the maximal ergodic theorem applied to g gives

$$\int_{E_a \cap E} f dP \geq aP(E_a \cap E)$$

for any invariant set E. Letting $-\infty < a < b < \infty$ be arbitrary and defining the invariant set

$$E_{a,b}(f) = \{liminf n^{-1} S_n < a < b < limsup n^{-1} S_n\}$$

we get from the above, noting that $E_{a,b} \subset E_b$ that

$$\int_{E_{a,b}(f)} f dP = \int_{E_{a,b}(f) \cap E_b} ddP \geq bP(E_{a,b}(f) \cap E_b) = bP(E_{a,b}(f))$$

Noting that

$$E_{a,b}(f) = \{liminf - n^{-1}S_n < -b < -a < limsup - n^{-1}S_n\} = E_{-b,-a}(-f)$$

gives us with f replaced by $-f$,

$$-\int_{E_{a,b}(f)} f dP = \int_{E_{-b,-a}(-f)} (-f) dP \geq -aP(E_{-b,-a}(-f)) = -aP(E_{a,b}(f))$$

and hence

$$bP(E_{a,b}(f)) \leq aP(E_{a,b}(f))$$

which implies since $b > a$ that

$$P(E_{a,b}(f)) = 0$$

proving thereby the ergodic theorem, ie,

$$liminf n^{-1} S_n = limsup n^{-1} S_n a.e P$$

or equivalently that

$$lim n^{-1} S_n$$

exists a.e. P. Note that we have to take a and b as rationals and form the countable union of the sets $E_{a,b}(f)$ over all rational $a < b$ to get zero for the probability of this union.

[9] Let ρ, σ be two quantum states in a finite dimensional Hilbert space $\mathcal{H}$. Prove that if $0 \leq s \leq 1$, then

$$Tr(\rho\{\rho \leq \sigma\}) \leq Tr(\rho^{1-s}\sigma^s)$$

hint:

$$\rho = \sigma - (\sigma - \rho) \geq \sigma - (\sigma - \rho)_+,$$

$$\sigma \geq \sigma - (\sigma - \rho)_+$$

Hence using operator monotonicity of $x \to x^s$ for $0 \leq s \leq 1$, we get

$$Tr(\rho^{1-s}\sigma^s) \geq Tr((\sigma - (\sigma - \rho)_+)^{1-s}(\sigma - (\sigma - \rho)_+)^s)$$

$$= Tr(\sigma - (\sigma - \rho)_+) \geq Tr((\sigma - (\sigma - \rho)_+)\{\rho > \sigma\})$$

$$= Tr(\sigma\{\rho > \sigma\})$$

Interchanging ρ and σ gives us the desired inequality.

[10] Let ρ, σ be two states in a given finite dimensional Hilbert space. Consider their spectral decompositions

$$\rho = \sum_i p(i)|e_i><e_i|, \sigma = \sum_i q(i)|f_i><f_i|$$

where

$$p(i), q(i) \geq 0, \sum_i p(i) = \sum_i q(i) = 1, <e_i|e_j>=<f_i|f_j>= \delta_{ij}$$

Express the relative entropy between these two states as a classical relative entropy between two probability distributions.

hint:

$$D(\rho|\sigma) = Tr(\rho.(log(\rho) - log(\sigma))) =$$

$$= \sum p(i)log(p(i) - \sum_{i,j} p(i)log(q(j))| < e_i|f_j > |^2$$

Define

$$P(i,j) = p(i)| < e_i|f_j > |^2, Q(i,j) = q(j)| < e_i|f_j > |^2$$

Then show using that

$$1 = \sum_i | < e_i|f_j > |^2 = \sum_j | < e_i|f_j > |^2$$

that $\{P(i,j)\}$ and $\{Q(i,j)\}$ are probability distributions and that

$$D(\rho|\sigma) = D(P|Q) = \sum_{i,j} P(i,j)log(P(i,j)/Q(i,j))$$

[11] In the theory of quantum binary hypothesis testing between two states ρ, σ, it is known that the optimal POVM test T that minimizes the error probability is attained at a PVM of the form

$$T = \{\rho > c\sigma\}$$

for some $c \in \mathbb{R}$. The corresponding minimum error can be expressed as

$$P(e) = P_+1Tr(\rho T) + P_2 Tr(\sigma(1 - T))$$

assuming the apriori probabilities of ρ and σ are respectively P_1 and $P_2 = 1 - P_1$. Now writing the spectral decompositions of ρ and σ as

$$\rho = \sum_i p(i)|e_i><e_i|, \sigma = \sum_i q(i)|f_i><f_i|$$

we get

$$Tr(\rho T) = \sum_i p(i) < e_i|T|e_i >$$

$$= \sum_i p(i) < e_i|T^2|e_i > = \sum_{i,j} p(i)| < e_i|T|f_j > |^2$$

$$= \sum_{i,j} p(j)| < f_i|T|e_j > |^2$$

$$Tr(\sigma(1-T)) = \sum_i q(i) < f_i|1-T|f_i >$$

$$= \sum_i q(i) < f_i|(1-T)^2|f_i > = \sum_{i,j} q(i)| < f_i|1-T|e_j > |^2$$

where we have used the fact that the optimum POVM T is actually a PVM which implies that $T^2 = T, (1-T)^2 = 1-T$. Thus the minimum error probability can be expressed as

$$P(e) = \sum_{i,j} (P_1 p(j)| < f_i|T|e_j > |^2 + P_2 q(i)| < f_i|1-T|e_j > |^2)$$

$$\geq \sum_{i,j} min(P_1 p(j), P_2 q(i))(| < f_i|T|e_j > |^2 + | < f_i|1-T|e_j > |^2)$$

Now, by the Schwarz inequality,

$$| < f_i|T|e_j > |^2 + | < f_i|1-T|e_j > |^2 \geq 2^{-1}(| < f_i|T|e_j > + < f_i|1-T|e_j > |^2)$$

$$= | < f_i|e_j > |^2/2$$

Hence,

$$P(e) \geq \sum_{i,j} (min(P_1 p(j), P_2 q(i))/2)| < f_i|e_j > |^2$$

$$= \sum_{i,j} min(P_1 P(i,j), P_2 Q(i,j))/2$$

where

$$P(i,j) = p(i)| < e_i|f_j > |^2, Q(i,j) = q(j)| < e_i|f_j > |^2$$

are two bivariate probability distributions. Now consider the problem of discriminating between the two classical probability distributions P, Q. The minimum error probability is

$$P(e) = min_{0 \leq t(i,j) \leq 1} \sum_{i,j} (P_1 P(i,j) t(i,j) + P_2 Q(i,j)(1 - t(i,j)))$$

$$= \sum_{i,j} [P_1 P(i,j) \chi_{P_1 P(i,j) \leq P_2 Q(i,j)} + P_2 Q(i,j) \chi_{P_2 Q(i,j) < P_1 P(i,j)}]$$

$$= P((i,j) : P(i,j)/Q(i,j) \leq P_2/P_1) + Q((i,j) : P(i,j)/Q(i,j) > P_2/P_1)$$

[12] Consider the problem of compression data taking into account distortion. Specifically, if the distortion in the encoding process is allowed to be present subject to the condition that it is smaller than a given threshold, then the number of compressed data bits per source bit can be reduced from $H(Q_1)$ to $minD(Q|Q_1 \times Q_2)$ where the minimum is taken over all joint distributions Q of the source symbol and the encoded symbol for which the first marginal is the given source distribution Q_1 and the distortion is smaller than the given allowable threshold D. Note that if X denotes the source symbol and Y the code symbol, then $Q(X = x, Y = y)$ is the joint distribution of (X, Y). If no distortion is permitted, the compressed number of bits per source symbol is $H(X)$ according to Shannon's coding theorem, but if a distortion of D is permitted, then the compressed number of bits per source bit reduces to

$$minI(X : Y) = min(H(X) - H(X|Y)) = H(X) - maxH(X|Y)$$

where $H(X|Y)$ is calculated using the joint distribution Q of (X, Y) and its maximum is over all joint distributions Q for which Q_1 is the given distribution of X and

$$\rho(Q) = \sum_{x,y} \rho(x, y)Q(x, y) \leq D$$

Explain this result and give all the proofs.

[13] Complete all the proofs in each of the following steps used in the proof of the direct part of the rate distortion theorem:
[1] Let A be a finite alphabet with a symbols and let Q_1 a given probability distribution on it. Let $C_n : A^n \to A^n$ be the deterministic compression code. By compression, we mean that generally $|C_n| < a^n$ where $|C_n|$ is the number of elements in $C_n(A^n)$. With ρ a metric on A, define the distortion of the code C_n as

$$\rho(C_n) = \mathbb{E}(n^{-1} \sum_{k=1}^{n} \rho(X_k, C_n(X)_k))$$

$$= n^{-1} \sum_{x \in A^n} \sum_{k=1}^{n} \rho(x_k, C_n(x)_k)Q_{1n}(x)$$

where Q_{1n} is the product distribution $Q_1^{\times n}$ on A^n, ie $(X_1, ..., X_n)$ are iid r.v's with each component having distribution Q_1. Optimum encoding for a given n means to select C_n so that $|C_n|$ is minimum subject to the condition that $\rho(C_n) \leq D$ for a given fixed positive real number D. Explain the meaning of such an encoding process.

[2] For a given distribution Q on $A \times A$ with first marginal Q_1 and second marginal Q_2 and a real number λ, define the distribution Q_λ on $A \times A$ by the formula

$$Q_\lambda(x, y) = \frac{exp(\lambda\rho(x, y))Q_1(x)Q_2(y)}{\sum_{y \in A} exp(\lambda\rho(x, y))Q_2(y)}$$

Show that Q_1 is the first marginal of Q_λ.

[3] With D denoting the relative entropy between two distributions, show that

$$0 \le D(Q|Q_\lambda) \le D(Q|Q_1 \times Q_2) + \Lambda_Q(\lambda) - \lambda\rho(Q)$$

where

$$\Lambda_Q(\lambda) = \sum_x Q_1(x)log(\sum_y exp(\lambda\rho(x,y))Q_2(y))$$

Deduce that

$$D(Q|Q_1 \times Q_2) \ge \Lambda_Q^*(\rho(Q))$$

where

$$\Lambda_Q^*(w) = sup_\lambda(\lambda w - \Lambda_Q(\lambda))$$

is the Legendre transform of Λ_Q.

[4] Define a random code $C_n : A^n \to A^n$ so that the elements in its range are $(Y(i,1), Y(i,2), ..., Y(i,n)), i = 1, 2, ..., k_n$ where $\{Y(i,j) : 1 \le i \le k_n, 1 \le j \le n\}$ are iid with distribution Q_2. Here, Q is a fixed probability distribution on $A \times A$ with first marginal Q_1. Define for $x \in A^n$, the set

$$S_n(x) = \{y \in A^n : n^{-1} \sum_{i=1}^{n} \rho(x_i, y_i) \le \rho(Q) + \delta\}$$

where δ is any fixed positive number. If $C_n(A^n) \cap S_n(x) \ne \phi$, then choose at any element $C_n(x)$ from this set. Note that $C_n(A^n)$ is a random set and hence $C_n(x)$ is actually one of the sequences $(Y(i,1), ..., Y(i,n)), i = 1, 2, ..., k_n$. Show that for any $x \in A^n$

$$n^{-1} \sum_{i=1}^{n} \rho(x_i, C_n(x)_i) \le \rho(Q) + \delta + \rho_{max}\chi_{C_n(A^n)\cap S_n(x)=\phi}$$

where

$$\rho_{max} = max(\rho(x,y) : x, y \in A)$$

is assumed to be finite.

[5] Let $X_1, X_2, ...$ be an ergodic process with probability distribution of X_1 being Q_1 and let $Y_1, Y_2, ...$ be iid with probability distribution Q_2 where as in the previous step, Q is a probability distribution on $A \times A$ with Q_1 and Q_2 as its two marginals. Define

$$Z_n(x) = n^{-1} \sum_{i=1}^{n} \rho(x_i, Y_i), x = (x_1, ..., x_n) \in A^n$$

Prove that if $(x_i)_{i=1}^\infty$ is a realization of the ergodic process (X_i), then

$$n^{-1}.log\mathbb{E}[exp(\lambda n Z_n)] = n^{-1} \sum_{i=1}^{n} log(\sum_{y \in A} exp(\lambda\rho(x_i, y))Q_2(y))$$

and that as $n \to \infty$, this converges for Q_1 a.e. x to $\Lambda_Q(\lambda)$. Then, apply the lower bound of the Gartner-Ellis large deviation theorem to deduce that

$$P(Z_n(x) \leq \rho(Q) + \delta) \geq exp(-n.\Lambda_Q^*(\rho(Q))) \geq exp(-nD(Q|Q_1 \times Q_2))$$

for Q_1 a.e. x when n is sufficiently large.

[6] Show that if $Y_1, Y_2, ..., Y(i,1), Y(i,2), ..., i = 1, 2, ...$ are iid with marginal distribution Q_2, then

$$P(S_n(x) \cap C_n(A^n)) = \phi) = P((Y(i,1), ..., Y(i,n)) \notin S_n(x), i = 1, 2, ..., k_n)$$

$$= P((Y_1, ...Y_{k_n}) \notin S_n(x))^{k_n}$$

$$\leq exp(-k_n.P(log(Z_n(x) \leq \rho(Q) + \delta)))$$

Choose

$$k_n = [exp(n(D(Q|Q_1 \times Q_2) + \delta))]$$

and deduce from the previous step and this step that

$$P(S_n(x) \cap C_n(A^n)) = \phi) \to 0, Q_1 a.e.x$$

Hence, conclude using step 4 that for this choice of $\{k_n\}$, we have

$$limpsup\rho(C_n) \leq \rho(Q) + \delta$$

Deduce from this the direct part of the rate distortion theorem.

[7] Give an intuitive proof of the direct part of Shannon's noisy coding theorem along the following lines. Let $X_n = (X(1), ..., X(n))$ and $Y_n = (Y(1), ..., Y(n))$ be respectively the input and output strings after time n. Assume that the $X(i)'s$ are iid with distribution p and that the channel is DMS with single symbol transition probability $\nu_x(y)$. Therefore, the output distribution is $q(y) = \sum_x p(x)\nu_x(y)$. Assume that n is sufficiently large.

[a] Show that the number of ϵ-entropy typical input sequences is approximately $2^{n(H(p))}$ when ϵ is small. Show that each such typical sequence has an approximate probability of $2^{-nH(p)}$.

[b] Show that the number of typical output sequences is approximately $2^{nH(q)}$ and that the probability of such each such typical sequence is approximately $2^{-nH(q)}$.

[c] The input-output sequence pair (X_n, Y_n) is said to be strongly jointly typical if X_n is typical, Y_n is typical and (X_n, Y_n) is jointly typical. Show that the probability that (X_n, Y_n) is jointly typical is $2^{-nH(X,Y)}$ and that the number of jointly typical sequences is $2^{nH(X,Y)}$.

[d] Let $(\tilde{X}_n, \tilde{Y}_n)$ be a joint input-output pair of n long sequences such that $\tilde{X}_n$ and $\tilde{Y}_n$ are independent but $\tilde{X}_n$ has the same distribution as X_n and $\tilde{Y}_n$ has the same distribution as Y_n. Show that the probability that $(\tilde{X}_n, \tilde{Y}_n) \in A_n$ is

$$2^{-nI(X,Y)} = 2^{-n(H(X)+H(Y)-H(X,Y))}$$

where A_n is the set of all jointly typical (X_n, Y_n). From this fact, deduce that there are roughly around $2^{nI(X,Y)}$ distinct typical input-output sequence pairs which we will come across before the joint sequence is also typical. Indeed, the number of i/o sequence pairs in which both the input and output sequences are typical is $2^{n(H(X)+H(Y))}$. The number of i/o pairs which are jointly typical is $2^{nH(X,Y)}$. Joint typicality is to be excluded since that results in decoding error. So if we remove the proportion of non-distinguishable input sequences (based on output data), we are left with $2^{nH(X)}.2^{nH(Y)}/2^{nH(X,Y)} = 2^{nI(X,Y)}$ distinguishable input sequences.

This means that there are in all $2^{nI(X,Y)}$ distinguishable input sequences. Another way to see this is that given that the output sequence is typical, the probability that the joint input-output sequence is also typical is $2^{nH(X|Y)}/2^{nH(Y)} = 2^{-nI(X,Y)}$.

Yet another way to see this is as follows. There are in all $2^{nH(X)}$ typical input sequences. For a given received output sequence, there are in all $2^{nH(X|Y)}$ input sequences for which the joint i/o pair is also typical. These joint typical pairs are to be excluded since they result in indistinguishability or error. Thus, the number of distinguishable/decodable input sequences from the output is $2^{nH(X)}/2^{nH(X|Y)} = 2^{nI(X,Y)}$.

Another way to see this is in terms of spheres: For any given input sequence x, all input sequences within a sphere V_x of volume $2^{nH(X|Y)}$ with centre x are jointly typical with the given output sequence y. The total volume of input typical sequences is $2^{nH(X)}$. All input sequences within the sphere V_x other than x will result in a decoding error since they are all jointly typical with y. Thus, the total number of sequences which do not result in decoding errors equals the total number of spheres V_x which are mutually disjoint and this number is evidently equal to the total volume $2^{nH(X)}$ of all the input typical sequences divided by the volume $2^{nH(X|Y)}$ of each sphere V_x. This ratio is $2^{n(H(X)-H(X|Y))} = 2^{nI(X,Y)}$.

[8] Let A, B denote respectively the input and output alphabets of a channel. Assume that the channel takes as input the infinite sequence $\mathbf{x} \in A^{\mathbb{Z}}$ and outputs the sequence $\mathbf{y} \in B^\infty$. Assume that the conditional probability $\nu(\mathbf{x}, [y_0, y_1, ..., y_n])$ only through $[x_{-m}, x_{-m+1}, ..., x_n]$, ie,

$$\nu(\mathbf{x}, [y_0, y_1, ..., y_n]) = \nu([x_{-m}, x_{-m+1}, ..., x_n], [y_0, y_1, ..., y_n])$$

for any non-negative integer n. Assume further that the channel is stationary, ie,

$$\nu([x_{-m}, ..., x_n], [y_0, ..., y_n]) = \nu([x_{-m+k}, ..., x_{n+k}], [y_k, ..., y_{n+k}])$$

for any integer k This means that the channel is stationary and has memory m. Assume further that the channel satisfies the m-independent property, ie, that conditioned on the input sequence, finite output sequences separated by a time lag greater than m are independent, ie,

$$\nu(\mathbf{x}, [y_{-p}, ..., y_0] \cap [y_{m+1}, ..., y_q]) =$$

$$\nu(\mathbf{x}, [y_{-p}, ..., y_0])\nu(\mathbf{x}, [y_{m+1}, ..., y_q])$$

for all $\mathbf{x}, \mathbf{y}$, non-negative p and $q > m$. Such a channel is said to be a stationary channel with memory m having the m-independence property. Prove then that if the input process to this channel is ergodic, then the joint input-output process is also ergodic.

[9] Prove Fano's inequality along the following lines:

Let X be the input r.v. and Y the output. From Y, we construct an estimate $\hat{X}$ of the input. Let $E = 1$ if $\hat{X} \neq X$ and $E = 0$ if $\hat{X} = X$. Assume that X takes values in the input alphabet A having a cardinality of a. Since the transformations $X \to Y \to \hat{X}$ form a Markov chain, it follows that that reversal $\hat{X} \to Y \to X$ also forms a Markov chain. Thus,

$$H(X|Y) = H(X|Y, \hat{X}) \leq H(X|\hat{X})$$

Further,

$$H(X|\hat{X}) \leq H(X, e|\hat{X}) = H(X|e, \hat{X}) + H(e|\hat{X})$$

$$= H(X|e = 1, \hat{X})P(e = 1) + H(e|\hat{X}) \leq P(e = 1)log_2(a - 1) + H(e)$$

$$\leq P(e = 1)log_2(a - 1) + 1$$

(since $e = \in \{0, 1\}$, it follows that $H(e) \leq log_2 2 = 1$). Noting that $P(e = 1) = P(X \neq \hat{X})$, we get on combining the above two inequalities that

$$P(X \neq \hat{X}) \geq (H(X|Y) - 1)/log_2(a - 1)$$

and this is called Fano's inequality. Use Fano's inequality to deduce the converse part of Shannon's noisy coding theorem.

[10] Prove the converse part of the Cq Shannon coding theorem using the following steps.

Let $\phi(i), i = 1, 2, ..., N$ be a code. Note that the size of the code is N. Let $s \leq 0$ and define

$$p_s = argmax_p I_s(p, W)$$

where

$$I_s(p, W) = min_\sigma D_s(p \otimes W | p \otimes \sigma)$$

where

$$p \otimes W = diag[p(x)W(x), x \in A], p \otimes \sigma = diag[p(x)sigma, x \in A]$$

Note that

$$D_s(p \otimes W | p \otimes \sigma) = log(Tr((p \otimes W)^{1-s}(p \otimes \sigma)^s)/(-s)$$

$$= (1/-s) \log sum_x Tr(p(x)W(x))^{1-s}(p(x)\sigma)^s)$$

$$= (1/-s) log \sum_x p(x)Tr(W(x)^{1-s}\sigma^s)$$

so that

$$lim_{s\to0} D_s(p \otimes W | p \otimes \sigma) =$$

$$\sum_x p(x)Tr(W(x)log(W(x)) - Tr(W_p.log(\sigma)))$$

where

$$W_p = \sum_x p(x)W_x$$

Note that this result can also be expressed as

$$lim_{s\to0} D_s(p \otimes W | p \otimes \sigma) = \sum_x p(x)D(W(x)|\sigma)$$

We define

$$f(t) = Tr(tW_x^{1-s} + (1-t)\sum_y p_s(y)W_y^{1-s})^{1/(1-s)}, t \geq 0$$

Note that

$$I_s(p, W) = (Tr(\sum_y p(y)W_y^{1-s})^{1/(1-s)})^{1-s}$$

and since by definition of p_s, we have

$$I_s(p_s, W) \geq I_s(p, W)\forall p,$$

it follows that

$$f(t) \leq f(0)\forall t \geq 0$$

Thus

$$f'(0) \leq 0$$

This gives

$$Tr[(W_x^{1-s} - \sum_y p_s(y)W_y^{1-s}).(\sum_y p_s(y)W_y^{1-s})^{s/(1-s)}] \leq 0$$

or equivalently,

$$Tr(W_x^{1-s}.(\sum_y p_s(y)W_y^{1-s})^{s/(1-s)}) \leq Tr(\sum_y p_s(y)W_y^{1-s})^{1/(1-s)}$$

Now define the state

$$\sigma_s = (\sum_y p_s(y)W_y^{1-s})^{1/(1-s)}/Tr(\sum_y p_s(y)W_y^{1-s})^{1/(1-s)}$$

Then we can express the above inequality as

$$Tr(W_x^{1-s}\sigma_s^s) \leq [Tr(\sum_y p_s(y)W_y^{1-s})^{1/(1-s)}]^{1-s}$$

Remark: Recall that

$$D_s(p \otimes W|p \otimes \sigma] = (1/-s)log\sum_x p(x)Tr(W_x^{1-s}\sigma^s)$$

$$= (1/-s)log(Tr(A\sigma^s))$$

where

$$A = \sum_x p(x)W_x^{1-s}$$

Note that we are all throughout assuming $s \leq 0$. Application of the reverse Holder inequality gives for all states σ

$$Tr(A\sigma^s) \geq (Tr(A^{1/(1-s)}))^{1-s}.(Tr(\sigma))^s = (Tr(A^{1/(1-s)}))^{1-s}$$

with equality iff σ is proportional to $A^{1/(1-s)}$, ie, iff

$$\sigma = A^{1/(1-s)}/Tr(A^{1/(1-s)})$$

Thus,

$$min_\sigma D_s(p \otimes W|p \otimes \sigma) =$$

$$(-1/s)[Tr(\sum_x p(x)W_x^{1-s})^{1/(1-s)}]^{1-s}$$

The maximum of this over all p is attained when $p = p_s$ and we denote the corresponding value of σ by σ_s. Thus,

$$\sigma_s = \frac{(\sum_x p_s(x)W_x^{1-s})^{1/(1-s)}}{Tr(\sum_x p_s(x)W_x^{1-s})^{1/(1-s)}}$$

Now let $\phi(i), i = 1, 2, ..., N$ be a code with $\phi(i) \in A^n$. Let $Y_i, i = 1, 2, ..., N$ be detection operators, ie, $Y_i \geq 0, \sum_i Y_i = 1$. Let $\epsilon(\phi)$ be the error probability corresponding to this code and detection operators. Then

$$1 - \epsilon(\phi) = N^{-1}\sum_{i=1}^N Tr(W_{\phi(i)}Y_i)$$

$$= N^{-1}Tr(S(\phi)T)$$

where

$$S(\phi) = N^{-1}diag[W_{\phi(i)}, i = 1, 2, ..., N], T = diag[Y_i, i = 1, 2, ..., N]$$

Also define for any state σ, the state

$$S(\sigma) = N^{-1}diag[\sigma_n, ..., \sigma_n] = N^{-1}.I_N \otimes \sigma_n$$

We observe that

$$Tr(S(\sigma_n)T) == N^{-1} \sum_i Tr(\sigma_n Y_i) = Tr(\sigma . N^{-1} . \sum_{i=1}^{N} Y_i)$$

$$= N^{-1} Tr(\sigma_n) = 1/N$$

where $\sigma_n = \sigma^{\otimes n}$. Then for $s \leq 0$ we have by monotonicity of quantum Renyi entropy, (The relative entropy between the pinched states cannot exceed the relative entropy between the original states)

$$N^{-s}(1 - \epsilon(\phi))^{1-s} = (Tr(S(\phi)T))^{1-s}(Tr(S(\sigma_{ns})T))^s$$

$$\leq (Tr(S(\phi)T))^{1-s}(Tr(S(\sigma_{ns})T))^s + (Tr(S(\phi)(1-T)))^{1-s}(Tr(S(\sigma_{ns})(1-T)))^s$$

$$\leq Tr(S(\phi)^{1-s}S(\sigma_{ns})^s)$$

Note that the notation $\sigma_{ns} = \sigma_s^{\otimes n}$ is being used. Now, we have

$$(Tr(S(\phi)^{1-s}S(\sigma_{ns})^s)) = (N^{-1} \sum_i Tr(W_{\phi(i)}^{1-s}\sigma_{ns}^s))$$

$$log(Tr(W_{\phi(i)}^{1-s}\sigma_{ns}^s)$$

$$= \sum_{l=1}^{n} log(Tr(W_{\phi_l(i)}^{1-s}\sigma_s^s))$$

$$= n.n^{-1} \sum_{l=1}^{n} log(Tr(W_{\phi_l(i)}^{1-s}\sigma_s^s))$$

$$\leq n.log(n^{-1}. \sum_{l=1}^{n} Tr(W_{\phi_l(i)}^{1-s}\sigma_s^s))$$

$$\leq n.log(Tr(\sum_y p_s(y)W_y^{1-s})^{1/(1-s)}]^{1-s})$$

$$= n(1-s).log(Tr(\sum_x p_s(x)W_x^{1-s})^{1/(1-s)})$$

Thus,

$$Tr(S(\phi)^{1-s}S(\sigma_{ns})^s) \leq (Tr(\sum_x p_s(x)W_x^{1-s})^{1/(1-s)})^{n(1-s)}$$

and hence

$$N^{-s}(1 - \epsilon(\phi))^{1-s} \leq (Tr(\sum_x p_s(x)W_x^{1-s})^{1/(1-s)})^{n(1-s)}$$

or equivalently,

$$1 - \epsilon(\phi) \leq exp((s/(1-s))log(N) + n.log(Tr(\sum_x p_s(x)W_x^{1-s})^{1/(1-s)}))$$

$$= exp((s/(1-s))log(N) + n.log(Tr(\sum_x p_s(x)W_x^{1-s})^{1/(1-s)}))$$

$$= exp(ns((1-s)log(N)/n + s^{-1}log(Tr(\sum_x p_s(x)W_x^{1-s})^{1/(1-s)}))) - - - (a)$$

where $s \leq 0$. Now, for any probability distribution $p(x)$ on A, we have

$$d/ds Tr[(\sum_x p(x)W_x^{1-s})^{1/(1-s)}] =$$

$$(1/(1-s)^2)Tr[(\sum_x p(x)W_x^{1-s})^{1/(1-s)}.log(\sum_x p(x)W_x^{1-s})]$$

$$-(1/(1-s))[Tr(\sum_x p(x)W_x^{1-s}log(W_x))].[Tr(\sum_x p(x)W_x^{1-s})^{s/(1-s)}]$$

and as $s \to 0$, this converges to

$$d/ds Tr[(\sum_x p(x)W_x^{1-s})^{1/(1-s)}]|_{s=0} =$$

$$\sum_x p(x)Tr(W_x.(log(W_p) - log(W_x))) = -\sum_x p(x)D(W_x|W_p)$$

$$= -(H(W_p) - \sum_p (x)H(W_x)) = -(H(Y) - H(Y|X)) = -I_p(X,Y)$$

where

$$W_p = \sum_x p(x)W_x$$

Write $N = N(n)$ and define

$$R = limsup_n log(N(n))/n, \phi = \phi_n$$

Also define

$$I_s(X,Y) = -(1/s(1-s))log(Tr(\sum_x p_s(x)W_x^{1-s})^{1/(1-s)}))$$

Then, we've just seen that

$$lim_{s \to 0}I_s(X,Y) = I_0(X,Y) \leq I(X,Y)$$

where $I_0(X,Y) = I_{p_0}(X,Y)$ and $I(X,Y) = sup_p I_p(X,Y)$. and

$$1 - \epsilon(\phi_n) \leq exp((ns/(1-s))(log(N(n))/n - I_s(X,Y)))$$

It thus follows that if

$$R > I(X,Y)$$

then by taking $s < 0$ and s sufficiently close to zero we would have

$$R > I_s(X,Y)$$

and then we would have

$$limsup_n(1 - \epsilon(\phi_n)) \leq limsup_n exp((ns/(1-s))(R - I_s(X,Y)) = 0$$

or equivalently,

$$liminf_n \epsilon(\phi_n) = 1$$

This proves the converse of the Cq coding theorem, namely, that if the rate of information transmission via any coding scheme is more that the Cq capacity $I(X,Y)$, then the asymptotic probability of error in decoding using any sequence of decoders will always converge to unity.

[11] Prove the direct part of the Cq coding theorem

Our proof is based on Shannon's random coding method. Consider the same scenario as above. Define

$$\pi_i = \{W_{\phi(i)} \geq 2NW_{p_n}\}, i = 1, 2, ..., N$$

where $N = N(n)$, $\phi(i) \in A^n$ with $\phi(i), i = 1, 2, ..., N$ being iid with probability distribution $p_n = p^{\otimes n}$ on A^n. Define the detection operators

$$Y(\phi(i)) = (\sum_{j=1}^{N} \pi)^{-1/2}.\pi_i.(\sum_{j=1}^{N} \pi_j)^{-1/2}, i = 1, 2, ..., N$$

Then $0 \leq Y_i \leq 1$ and $\sum_{i=1}^{N} Y_i = 1$. Write

$$S = \pi_i, T = \sum_{j \neq i} \pi_j$$

Then,

$$Y(\phi(i)) = (S + T)^{-1/2}S(S + T)^{-1/2}, 0 \leq S \leq 1, T \geq 0$$

and hence we can use the inequality

$$1 - (S + T)^{-1/2}S(S + T)^{-1/2} \leq 2 - 2S + 4T$$

Then,

$$1 - Y(\phi(i)) =\leq 2 - 2\pi_i + 4 \sum_{j:j \neq i} \pi_j$$

and hence

$$\epsilon(\phi) = N^{-1} \sum_{i=1}^{N} Tr(W_{\phi(i)}(1 - Y(\phi(i))))$$

$$\leq N^{-1} \sum_{i=1}^{N} Tr(W(\phi(i))(2 - 2\pi_i) + 4N^{-1} \sum_{i \neq j} Tr(W(\phi(i))\pi_j)$$

Note that since for $i \neq j$, $\phi(i)$ and $\phi(j)$ are independent random variables in A^n, it follows that $W(\phi(i))$ and π_j are also independent operator valued random variables. Thus,

$$\mathbb{E}(\epsilon(\phi)) \leq (2/N) \sum_{i=1}^{N} \mathbb{E}(Tr(W(\phi(i))\{W(\phi(i)) \leq 2NW_{p_n}\}))$$

$$+(4/N) \sum_{i \neq j, i, j = 1, 2, \ldots, N} Tr(\mathbb{E}(W(\phi(i))).\mathbb{E}\{W(\phi(j)) > 2NW_{p_n}\})$$

$$= 2 \sum_{x_n \in A^n} p_n(x_n) Tr(W(x_n)\{W(x_n) \leq 2N.W(p_n)\})$$

$$+4(N-1) \sum_{x_n, y_n \in A^n} p_n(x_n) p_n(y_n) Tr(W(x_n)\{W(y_n) > 2NW(p_n)\})$$

$$= (2/N) \sum_{x_n} p_n(x_n) Tr(W(x_n)\{W(x_n) \leq 2N.W(p_n)\})$$

$$+4(N-1) \sum_{y_n \in A^n} p_n(y_n) Tr(W(p_n)\{W(y_n) > 2NW(p_n)\})$$

$$\leq 2 \sum_{x_n} p_n(x_n) Tr(W(x_n)^{1-s}(2NW(p_n))^s)$$

$$+4N \sum_{y_n} p_n(y_n) Tr(W(y_n)^{1-s} W(p_n)^s (2N)^{s-1})$$

for $0 \leq s \leq 1$. This gives

$$\mathbb{E}(\epsilon(\phi_n)) \leq$$

$$2^{s+2} N^s \sum_{x_n \in A^n} p_n(x_n) Tr(W(x_n)^{1-s} W(p_n)^s)$$

$$= 2^{s+2} N^s [\sum_{x \in A} p(x) Tr(W(x)^{1-s} W(p)^s)]^n$$

$$= 2^{s+2}.exp(sn(log(N(n)/n + s^{-1}.log(\sum_x p(x) Tr(W(x)^{1-s} W(p)^s))))$$

[12] Use Fano's inequality to give an intuitive proof of the converse of Shannon's noisy coding theorem.

hint: Let W be the message r.v. assumed to take one of 2^{nR} values. For each message w, we assign a code-word X_n in A^n. Thus, the code rate is $nR/n = R$. The channel is DMS and it takes the n-long string X_n as input and outputs the n-long string Y_n. Let $\hat{W}$ denote the decoded message. We are assuming that the input and output alphabets are $\{0, 1\}$. Then, the transformation $W \rightarrow X_n \rightarrow$

$Y_n \to \hat{W}$ is a Markov chain and assuming W to have the uniform distribution over the 2^{nR} messages, we have

$$nR = H(W) = H(W|\hat{W}) + I(W, \hat{W}) \leq 1 + nP(W \neq \hat{W}) + I(X_n, Y_n)$$

$$\leq 1 + nP(W \neq \hat{W}) + nI(X, Y)$$

The probability of decoding error for the message W is $P(W \neq \hat{W})$ and if this converges to zero as $n \to \infty$, then the above equation implies that

$$R \leq lim_{n \to \infty}(n^{-1} + P(W \neq \hat{W}) + I(X, Y)) = I(X, Y)$$

and this completes the proof of the converse part of Shannon's noisy coding theorem.

Remark: Fano's inequality states that

$$P(W \neq \hat{W}) \geq (H(W|\hat{W}) - 1)/log(2^{nR} - 1)$$

which in turn implies that

$$H(W|\hat{W}) \leq 1 + nR.P(W \neq \hat{W}) \leq 1 + n.P(W \neq \hat{W})$$

[13] Let $W \to X \to Y \to V$ be a Markov chain. Then, show that $V \to Y \to X \to W$ is also a Markov chain, ie, time reversal of a Markov chain is also a Markov chain. Show further that

$$I(W, V) \leq I(X, Y)$$

[14] Let $0 \leq s \leq 1$. The aim of this exercise is to show that if $\mathcal{H} = \mathcal{H}_1 \otimes \mathcal{H}_2$ and if K is the partial trace operation over $\mathcal{H}_2$, then for any two states ρ, σ in $\mathcal{H}$ and $0 \leq s \leq 1$, we have

$$Tr(\rho^{1-s}\sigma^s) \leq Tr(K(\rho)^{1-s}K(\sigma)^s)$$

Step 1: Show that $x \to x^s = f(x)$ on $\mathbb{R}_+$ is operator concave for $0 \leq s \leq 1$.

Step 2: For an operator concave function f on $\mathbb{R}_+$, show that $f(ZAZ^*) \geq Zf(A)Z^*$ for any positive A with Z an arbitrary operator.

Step 3: Let ρ, σ be any two states in $\mathcal{H}$. Define the linear operator $\Delta_{\rho,\sigma}$ on $\mathcal{H}$ by

$$\Delta_{\rho,\sigma}(X) = \sigma^{-1}.X\rho$$

Also define an inner product $< .,. >_\sigma$ on $\mathcal{H}$ by

$$< X, Y >_\sigma = Tr(X^* \sigma Y)$$

Show that $\Delta_{\rho,\sigma}$ is a positive operator in $\mathcal{H}$ w.r.t the inner product $< .,. >_\sigma$.

Step 4: Define the operator $K_\sigma : \mathcal{L}(\mathcal{H}) \to \mathcal{L}(\mathcal{H}_1)$, by

$$K_\sigma(X) = K(\sigma)^{-1} K(\sigma.X)$$

Show that

$$< Y, K_\sigma(X) >_{K(\sigma)} = Tr(Y^* K(\sigma X)) = Tr((Y \otimes I_2)^* \sigma X)$$

$$= < Y \times I_2, X >_\sigma$$

for any $X \in \mathcal{L}(\mathcal{H})$ and $Y \in \mathcal{L}(\mathcal{H})$. Hence deduce that the dual K_σ^* of K_σ w.r.t the inner products $< .,. >_{K(\sigma)}$ and $< .,. >_\sigma$ on $\mathcal{H}_1$ and $\mathcal{H}$ is given by

$$K_\sigma^*(Y) = Y \times I_2, Y \in \mathcal{L}(\mathcal{H}_1)$$

Question paper on Information Theory

[1] Let A and B be two finite alphabets with cardinalities a and b respectively. What do you understand by a code $f : A \to S(B)$ where $S(B)$ is the infinite set of all strings in B ? When is such a code said to satisfy the no-prefix condition ? When is such a code said to be uniquely decipherable ? Show that a code satisfying the no prefix condition is also uniquely decipherable and by means of an example show that a uniquely decipherable code need not satisfy the no prefix condition. Prove that f satisfies the no prefix condition if and only if $\sum_{x \in A} b^{-l(f(x))} \leq 1$ where $l(f(x))$ is the length of the string $f(x) \in S(B)$. Illustrate your answer by drawing a tree diagram for a no-prefix code.

[2] Show that if $f : A \to S(B)$ is a code satisfying the no prefix condition, then

$$\bar{l}(f, \mu) \geq H(\mu)/log_2(b)$$

where $H(\mu) = -\sum_{x \in A} \mu(x).log_2(\mu(x))$ is the Shannon entropy of the probability distribtution μ of the source alphabet $x \in A$ and $l(f, \mu) = \sum_{x \in A} l(f(x))\mu(x)$ is the average codeword length of f. Show in addition that there exists a code $f : A \to S(B)$ satisfying the no prefix condition for which

$$\bar{l}(f, \mu) \leq 1 + H(\mu)/log_2 b$$

Show that such a code can be constructed by choosing $l(f(x))$ so that

$$l(f(x)) - 1 < -log_b \mu(x) \leq l(f(x)), x \in A$$

Illustrate the construction of such a code for the probabilities $\mu(1) = 0.5, \mu(2) = 0.24, \mu(3) = 0.1, \mu(4) = 0.25$ when A has four alphabets and $B = \{0,1\}$, ie, $b = 2$. Draw a tree diagram for this code.

hint: Make use of Kraft's inequality and the fact that if p, q are any two probability distributions on the same set X, then $\sum_{x \in X} p(x) log(p(x)/q(x)) \geq 0$.

[3] Show that if if $f : A^n \to S(B)$ is a code satisfying the no prefix condition and μ is the source probability distribution on A so that the source probability distribution on A^n is the product distribution, ie the probability of $(x_1...x_n) \in A^n$ occurring if $\mu(x_1)...\mu(x_n)$, and then if f is the length of a code having minimum average length $\bar{l}_n(f,\mu) - \sum_{x_1,...,x_n \in A} l(f(x_1...x_n))\mu(x_1)...\mu(x_n)$, then the minimum average codeword length $\bar{l}_n(f,\mu)/n$ per source symbol satisfies the inequalities

$$H(\mu)/log_2 b \leq \bar{l}_n(f,\mu)/n \leq H(\mu)/log_2 b + 1/n$$

Deduce that by choosing n sufficiently large and encoding blocks of n-long source sequences rather than single source words, the average code word length per source symbol can be made arbitrarily close to $H(\mu)/log_2 b$ with the code satisfying the no prefix condition.

[4] Let X, Y be two random variables. Show that

$$I(X,Y) = H(X) + H(Y) - H(X,Y) = H(X) - H(X|Y) = H(Y) - H(Y|X) \geq 0$$

What do you understand by an entropy typical sequence $T_E(n, \mu, \delta)$ for a discrete memoryless source with probability distribution μ. Show that

$$\mu^{(n)}(T_E(n, \mu, \delta)) \geq 1 - \sigma^2/n\delta^2$$

where σ^2 is a constant equal to $Var(log(\mu(X)))$. Hence, deduce the direct part of the Shannon noiseless coding theorem using the definition of entropy typical sequences and Chebyshev's inequality:

$$lim_{\delta \to 0} lim_{n \to \infty} n^{-1} log_2 N(T_E(n, \mu, \delta)) = H(\mu)$$

Specifically show that

$$H(\mu) - \delta) \leq liminf_n \frac{log_2(N(T_E(n, \mu, \delta)))}{n} \leq limsup_n \frac{N(T_E(n, \mu, \delta))}{n} \leq H(\mu + \delta)$$

for any $\delta > 0$.

hint: Define

$$T_E(n, \mu, \delta) = \{(x_1, ..., x_n) \in A^n : |n^{-1} \sum_{i=1}^n log(\mu(x_i)) + H(\mu)| < \delta\}$$

Also deduce the converse of the Shannon noiseless conding theorem, namely that if $E_n \subset A^n, n = 1, 2, ...$ is such that $\mu^{(n)}(E_n) \to 1$ as $n \to \infty$, then

$$liminf_n n^{-1} log(N(E_n) \geq H(\mu)$$

Give the physical interpretation of these two results from the viewpoint of noiseless data compression.

[5] Use the lower bound of the large deviation principle to show that if A is a source alphabet and the source is ergodic, then for any $\delta, D > 0$ there exists a sequence of codes $C_n : A^n \to A^n$ such that

$$limsup_n \frac{log(|C_n(A^n)|}{n} \leq R(D) + \delta$$

and simultaneously achieve an average distortion

$$limsup_n n^{-1} \mathbb{E}_{Q_1^{(n)}} \sum_{k=1}^{n} \rho(x_k, C_n(x)_k) \leq D + \delta$$

where

$$R(D) = inf_Q\{H(Q|Q_1 \times Q_2) : \rho(Q) \leq D\}$$

is the rate distortion function and the infimum is taken over all probability distributions Q on $A \times A$ for which the first marginal of Q is the given source distribution Q_1. $Q_1^{(n)}$ is the distribution of $(x_1, ... x_n)$ which as $n \to \infty$ is assumed to be ergodic. Here,

$$\rho(Q) = \sum_{x,y \in A} Q(x, y) \rho(x, y)$$

is the average distortion induced by Q. Make use of the large deviation principle in the form that for large n

$$n^{-1} log P(Z_n(x) \leq \rho(Q) + \delta) \geq -\Lambda_Q^*(\rho(Q)), a.e. x, Q_1$$

where

$$\Lambda_Q^*(y) = sup_\lambda(\lambda y - \Lambda_Q(\lambda))$$

where

$$Z_n(x) = n^{-1} \sum_{k=1}^{n} \rho(x_k, Y_k)$$

and

$$\Lambda_Q(\lambda) = \sum_x Q_1(x) log(\sum_y exp(\lambda \rho(x, y)) Q_2(y))$$

To arrive at this, make use of the ergodicity of the input string $x_1, x_2, ...$ in the form

$$lim_n n^{-1} \sum_{k=1}^{n} F(x_k) = \sum_{x \in A} F(x) Q_1(x), a.e$$

with $Y_1, ..., Y_n$ being iid with distribution Q_2, the second marginal of Q. You will also have to make use of the identity

$$0 \le H(Q|Q_\lambda)$$

where Q_λ is the distribution on $A \times A$ defined by

$$Q_\lambda(x, y) = exp(\lambda \rho(x, y))Q_1(x)Q_2(y)/ \sum_z exp(\lambda \rho(x, z))Q_2(z)$$

Interpret this result from the standpoint of data compression in the sense that is a minimum amount of distortion is permitted during data compression, then we can compress the data more that $H(X) = H(Q_1)$ per source symbol as given by Shannon's noiseless coding theorem.

[6] Give an intuitive proof of Shannon's noisy coding theorem along the following lines: The number of input typical sequences of length n is $2^{nH(X)}$. Given the received output sequence Y, the number of input typical sequences is $2^{nH(X|Y)}$. If an input sequence X_n of length n is transmitted and and the received sequence is Y_n, then a decoding error occurs if either (X_n, Y_n) is not jointly typical or else if some other input sequence X'_n and Y_n are jointly typical. Hence, argue that for a given transmitted input and the given received output, the number of decoding errors is $2^{nH(X|Y)}$. Deduce thereby that the number of input sequences that are distinguishable from the given output sequence is the ratio $2^{nH(Y)}/2^{nH(X|Y)} = 2^{n(H(Y)-H(X|Y))}$.

[7] Write down the Huffman code for a source probability distribution $\{0.4, 0.3, 0.2, 0.1\}$ and prove the optimality (minimum average length of the Huffman code.

More problems in information theory

[1] Let $\rho_1(i), \rho_2(i), i = 1, 2, ..., N$ be states in a Hilbert spaces $\mathcal{H}_1$ and $\mathcal{H}_2$ respectively. Define

$$\rho(i) = \rho_1(i) \otimes \rho_2(i), 1 \le i \le N$$

and note that $\rho(i)$ is a state in $\mathcal{H} = mathcal H_1 \otimes \mathcal{H}_2$. Let $p = \{p(i) : 1 \le i \le N\}$ be a probability distribution on $\{1, 2, ..., N\}$. Consider the inequality

$$D(\sum_i p_i \rho(i) | \rho_{mix} \otimes \sum_i p(i)\rho_2(i)) \ge 0$$

Show that this inequality yields

$$-H(\sum_i p(i)\rho_1(i) \otimes \rho_2(i)) + log(d_1) + H(\sum_i p(i)\rho_2(i)) \ge 0$$

[2] Let $M = \{M(i) : i = 1, 2, ..., a\}$ be a POVM. Let ρ, σ be two states. Define the probability distributions

$$P_\rho^M(i) = Tr(\rho.M(i)), P_\sigma^M(i) = Tr(\sigma.M(i)), 1 \le i \le a$$

Show that
$$D(\rho|\sigma) \geq D(P_\rho^M | P_\sigma^M)$$

by considering the TPCP map K defined by

$$K(\rho) = \sum_i Tr(\rho M(i))|e_i><e_i| = \sum_i P_\rho^M(i)|e_i><e_i|$$

where $\{|e_i>\}$ is an onb for the Hilbert space and using the fact that the relative entropy is monotone, ie,

$$D(\rho|\sigma) \geq D(K(\rho)|K(\sigma))$$

Remark:First prove that K is a TPCP map and determine its Stinespring representation.

We can write

$$Tr(\rho M(i)) = \sum_{j,k} <e_j|\rho|e_k><e_k|M(i)|e_j>$$

$$= \sum_{j,k} M_i(k,j) <e_j|\rho|e_k>$$

where

$$M_i(k,j) = <e_k|M(i)|e_j>$$

Thus,

$$K(\rho) = \sum_{ijk} M_i(k,j) <e_j|\rho|e_k> |e_i><e_i|$$

$$= \sum_{jk} M(k,j) <e_j|\rho|e_k>$$

where

$$M(k,j) = \sum_i M_i(k,j)|e_i><e_i|$$

Consider the block structured matrix

$$M = ((M(k,j)))$$

Each block $M(k,j)$ here is an $n \times n$ block and we have

$$M(k,j)^* = M(j,k)$$

and in particular, M is a self-adjoint $n^2 \times n^2$ matrix consisting of $n \times n$ blocks, each block being of size $n \times n$. We have the spectral decomposition

$$M = UDU^*$$

where U is $n^2 \times n^2$ unitary and D is $n^2 \times n^2$ diagonal with positive diagonal entries. Equivalently, in $n \times n$ block structured form, the spectral decomposition of the self-adjoint matrix $((M(k, j)^*))$ can be expressed as

$$M(k, j)^* = M(j, k) = \sum_l U(k, l)D(l)U)(j, l)^*$$

where $U(k, l)$ is $n \times n$ and $D(l)$ is $n \times n$ diagonal. Equivalently,

$$M(k, j) = \sum_l U(j, l)D(l)U(k, l)^*$$

Then,

$$K(\rho) = \sum_{kjl} U(j, l)D(l)U(k, l)^* < e_j|\rho|e_k >$$

Writing

$$E(l) = \sum_l U(j, l)\sqrt{D(l)} \otimes < e_j|$$

we can express this as

$$K(\rho) = \sum_l Tr_2(E(l)(I_n \otimes \rho)E(l)^*)$$

which easily yields the desired Stinespring representation of K.

[3] Consider the states

$$\rho = \sum_i p(i)\rho_1(i) \otimes \rho_2(i), \sigma = \rho_{mix} \otimes \sum_i p(i)\rho_2(i)$$

Let K denote partial trace w.r.t the second Hilbert space. Then, K is a quantum operation and hence we have by monotonicity of quantum relative entropy that

$$D(\rho|\sigma) \geq D(K(\rho)|K(\sigma))$$

Now,

$$D(\rho|\sigma) = Tr(\rho.(log(\rho) - log(\sigma))) =$$

$$-H(\rho) + log(d_1) - Tr(\sum_i p(i)\rho_2(i)log(\sum_i p(i)\rho_2(i)))$$

$$= -H(\rho) + log(d_1) + H(p) + \sum_i p(i)H(\rho_2(i)),$$

$$K(\rho) = \sum_i p(i)\rho_1(i), K(\sigma) = \rho_{mix}$$

so that

$$D(K(\rho)|K(\sigma)) = -H(\sum_i p(i)\rho_1(i)) + log(d_1)$$

and thus we obtain the inequality

$$H(p) + \sum_i p(i)H(\rho_2(i)) + H(\sum_i p(i)\rho_1(i)) \geq H(\sum_i p(i)\rho_1(i) \otimes \rho_2(i))$$

[4] Consider the state

$$\rho = \sum_i p(i)|e_i >< e_i| \otimes W(i)$$

where p is a probability distribution, $\{|e_i >\}$ is an onb for the first Hilbert space and $W(i)$ are states in the second Hilbert space. Define

$$I(p,W) = H(\rho_1) + H(\rho_2) - H(\rho)$$

where

$$\rho_1 = Tr_2(\rho), \rho_2 = Tr_1(\rho)$$

Evaluate $I(p,W)$ and show that it is non-negative.

If ρ is any state on the tensor product of two Hilbert spaces and $\rho_1 = Tr_2(\rho), \rho_2 = Tr_1(\rho)$, then

$$H(\rho_1) + H(\rho_2) - H(\rho) = D(\rho|\rho_1 \otimes \rho_2) \geq 0$$

because the relative entropy between any two states is always non-negative. In the special case of this problem, we have

$$\rho_1 = \sum_i p(i)|e_i >< e_i|, \rho_2 = \sum_i p(i)W(i)$$

and

$$H(\rho_1) = H(p), H(\rho_2) = H(\sum_i p(i)W(i)),$$

$$H(\rho) = -Tr(\rho.log(\rho)) = -Tr(\sum_i p(i)W(i)log(p(i)W(i)))$$

$$= H(p) + \sum_i p(i)H(W(i))$$

so we get

$$I(p,W) = H(\sum_i p(i)W(i)) - \sum_i p(i)H(W(i)) \geq 0$$

It should be noted that the supremum of $I(p,W)$ over all p is the capacity of the Cq channel defined by $i \rightarrow W(i)$.

[5] Prove that if $0 \leq S \leq I$ and $T \geq 0$, then

$$I - (S+T)^{-1/2}S(S+T)^{-1/2} \leq 2I - 2S + 4T$$

hint: Define the matrices

$$A = \sqrt{T}, B = \sqrt{T}((T+S)^{-1/2} - I)$$

and consider the inequality

$$0 \leq (A-B)^*(A-B) = A^*A + B^*B - A^*B - B^*A$$

leading to

$$A^*A + B^*B + A^*B + B^*A \leq 2(A^*A + B^*B)$$

Show that the lhs of this inequality is

$$(T+S)^{-1/2}T(T+S)^{-1/2} = I - (T+S)^{-1/2}S(T+S)^{-1/2}$$

and that the rhs is

$$2T + 2(R^{-1/2} - 1)T(R^{-1/2} - 1)$$
$$\leq 2T + 2(R^{-1/2} - 1)R(R^{-1/2} - 1)$$
$$= 2T + 2 + 2R - 4R^{1/2} = 4T + 2 + 2S = 2 + 4T + 2S - 4R^{1/2}$$

where

$$R = T + S$$

Show that $T \geq 0, 0 \leq S \leq 1$ together imply

$$R^{1/2} = (T+S)^{1/2} \geq S^{1/2} \geq S$$

and therefore the rhs is $\leq$

$$2 + 4T + 2S - 4S = 2 - 2S + 4T$$

thereby completing the proof.

[6] A proof of Shannon's rate distortion theorem without the use of large deviation theory. Let $x \in A^n$ with A the source alphabet. Thus, the source space is A^n and it consists of a^n strings. The code space is a subset of A^n. We denote this code space by C. Typically, $|C| << a^n$. We write $\hat{x} \in C$ for a code word. Assume that the code space C consists of 2^{nR} strings. Thus, for compression, we wish that $R < 1$. We select a joint distribution $p(x, \hat{x})$ for the source and code strings in such a way that

$$p(x, \hat{x}) = \Pi_{i=1}^n p_0(x_i, \hat{x}_i)$$

where

$$x = (x_i)_{i=1}^n, \hat{x} = (\hat{x}_i)_{i=1}^n$$

In other words, the joint pdf of the source and code strings is selected so that the joint components are iid with joint distribution p_0. We say that $(x, \hat{x})$ is (n, ϵ)-distortion typical if

$$\left| n^{-1} \sum_{k=1}^n d(x_k, \hat{x}_k) - D \right| < \epsilon,$$

$$|n^{-1}log(p_1(x)) + H(p_{01})| < \epsilon,$$

$$|n^{-1}log(p_2(\hat{x})) - H(p_{12})| < \epsilon,$$

$$|n^{-1}log(p(x, \hat{x})) - H(p_0)| < \epsilon$$

where $p_1(x) = \sum_{\hat{x}} p(x, \hat{x})$ and $p_2(\hat{x}) = \sum_x p(x, \hat{x})$ are respectively the distributions of the strings x and $\hat{x}$ and p_{01} is the distribution of x_k for any k while p_{02} is the distribution of $\hat{x}_k$ for any k. Note that since $(x_k, \hat{x}_k), k = 1, 2, ..., n$ are iid pairs with distribution p_0, it follows that $x_k, k = 1, 2, ..., n$ are iid with distribution p_{01} and $\hat{x}_k, k = 1, 2, ..., n$ are iid with distribution p_{02}. In the above definitions,

$$D = \mathbb{E}d(x_k, \hat{x}_k) = \sum_{x_1, \hat{x}_1} p_0(x_1, \hat{x}_1)d(x_1, \hat{x}_1)$$

is the average distortion per symbol. Let $A(d, n, \epsilon)$ denote the set of all (n, ϵ) distortion typical pairs $(x, \hat{x})$.

We now generate a random code $C \subset A^n$ consisting of 2^{nR} strings, ie for each $x \in A^n$, we randomly assign a codeword $\hat{x} = \hat{x}(w) \in A^n$ where $w = 1, 2, ..., 2^{nR}$ with the distribution of $\hat{x}$ being $p_2(\hat{x}) = \Pi p_{02}(\hat{x}_k)$, namely the second marginal of p. We next observe that by the weak law of large numbers,

$$p(\hat{x}|x)/p_2(\hat{x}) = p(x, \hat{x})/p_1(x)p_2(\hat{x})$$

$$\leq 2^{-n(H(x_1, \hat{x}_1) - \epsilon)}/2^{-n(H(x_1) + \epsilon)}.2^{-n(H(\hat{x}_1) + \epsilon)}$$

$$= 2^{n(I(x_1, \hat{x}_1) + 3\epsilon)}$$

whenever $(x, \hat{x}) \in A(d, n, \epsilon)$. Note that under p, the probability of $A(d, n, \epsilon)$ converges to zero as $n \to \infty$. In other words, in the limit of large sized sequences, almost all source-code sequence pairs are jointly typical. Note also that we can write

$$H(x_1) = H(p_{01}), H(\hat{x}_1) = H(p_{02}), H(x_1, \hat{x}_1) = H(p_0)$$

We now define $K(x, \hat{x}) = \chi_{A(d,n,\epsilon)}(x, \hat{x})$, ie $K(x, \hat{x}) = 1$ if $(x, \hat{x})$ is (n, ϵ)-distortion typical and zero otherwise. We define $J(C)$ to be the set of all input sequences $x \in A^n$ for which there exists an $\hat{x} \in C$ for which $(x, \hat{x})$ is (n, ϵ)-distortion typical, ie,

$$J(C) = \{x | (x, \hat{x}) \in A(d, n, \epsilon) for some \hat{x} \in C\}$$

For a given source string $x \in A^n$, with the random code C constructed as above, we assign that codeword $\hat{x}(w)$ to x for which $(x, \hat{x}(w))$ is distortion typical. if there are two or more such $w's$ in $1, 2, ..., 2^{nR}$, then we assign that $\hat{x}(w)$ for which w is a minimum. In this way each $x \in A^n$ is encoded into a unique codeword $\hat{x}(w)$. Then, it is easy to see that with

$$d_n(x, \hat{x}) = n^{-1} \sum_{k=1}^{n} d(x_k, \hat{x}_k) \leq D + \epsilon + d_{max}\chi_{A(d,n,\epsilon)^c}(x, \hat{x})$$

because if $(x, \hat{x}) \in A(d, n, \epsilon)$, then $d_n(x, \hat{x}) \leq D + \epsilon$ from the definition of $A(d, n, \epsilon)$, while if $(x, \hat{x}) \in A(d, n\epsilon)^c, d_n(x, \hat{x}) \leq d_{max}$. Now we take the expected value on both sides of the above inequality over all source words x and the corresponding codewords $\hat{x} \in C$ and then over all random codes C described as above. For such an expectation, we find that

$$\mathbb{E}(d_n(x, \hat{x})) \leq D + \epsilon + d_{max} P(e)$$

where

$$P(e) = \sum_C P(C)(P(A(d, n, \epsilon)^c)) = \sum_C P(C)P((x, \hat{x}) \notin A(d, n, \epsilon))$$

$$= \sum_C P(C) \sum_{x \notin J(C)} p_1(x)$$

Remark: The event $x \notin J(C)$ occurs iff $(x, \hat{x}) \notin A(d, n, \epsilon)$ for all $\hat{x} \in C$. Now,

$$\sum_C P(C) \sum_{x \notin J(C)} p_1(x) = \sum_x p_1(x) \sum_{C : x \notin C} P(C)$$

$$= \sum_x p_1(x)(1 - \sum_{\hat{x}} p_2(\hat{x}) K(x, \hat{x}))^{2^{nR}}$$

Remark: Writing the elements of the random code C as $w_1, w_2, ..., w_M, M = 2^{nR}$, it follows that

$$\sum_{C : x \notin C} P(C) = \sum_{w_1, ..., w_M : (x, w_k) \notin A(d, n, \epsilon), k = 1, 2, ..., M} p_2(w_1)...p_2(w_M)$$

$$= (\sum_{w_1 : (x, w_1) \notin A(d, n, \epsilon)} p_2(w_1))^M$$

$$= (1 - \sum_{\hat{x}} p_2(\hat{x}) K(x, \hat{x}))^M$$

Remark: More precisely, if we compute the average of $d(x, \hat{x})$ over all source words x and all codes C, then this will be a sum of two terms. The first will be a sum over all x for which there exists an $\hat{x}$ such that $(x, \hat{x})$ is distortion typical. For all such x, we have that $d_n(x, \hat{x}) \leq D + \epsilon$. Note that by construction, for all such x, $\hat{x} \in C$ is uniquely determined. The second sum will be a sum over all x for which there does not exist any $\hat{x} \in C$ for which $(x, \hat{x})$ is distortion typical. For all such $x's$, we assign a single codeword $\hat{x}$ arbitrarily. This sum will be dominated by $d_{max} P(e)$. Further, using the inequality proved above, namely,

$$p_2(\hat{x}) \geq 2^{-n(I+3\epsilon)} p(\hat{x}|x)$$

we get

$$P(e) \leq \sum_x p_1(x)(1 - 2^{-n(I+3\epsilon)} \sum_{\hat{x}} p(\hat{x}|x) K(x, \hat{x}))^M$$

$$\leq \sum_x p_1(x)[(1 - \sum_{\hat{x}} p(\hat{x}|x)K(x,\hat{x})) + exp(-2^{-n(I+3\epsilon)}M)\sum_{\hat{x}} p(\hat{x}|x)K(x,\hat{x})]$$

$$= \sum_{x,\hat{x}} p(x,\hat{x})(1 - K(x,\hat{x})) + exp(-2^{n(R-I-3\epsilon)})\sum_{x,\hat{x}} p(x,\hat{x})K(x,\hat{x})$$

$$= P((x,\hat{x}) \in A(d,n,\epsilon)^c) + exp(-2^{n(R-I-3\epsilon)}).P((x,\hat{x}) \in A(d,n,\epsilon))$$

where the above probabilities are now computed using p. It is clear then that if $R > I + 3\epsilon$, $P(e) \to 0$ as $n \to \infty$, since by the weak law of large numbers, $P((x,\hat{x}) \in A(d,n,\epsilon)^c) \to 0$.

[1] Given a wide sense stationary random process $x[n], n \in \mathbb{Z}$ with zero mean and autocorrelation $R(\tau) = \mathbb{E}(x[n+\tau]x[n])$, derive a formula for its pth order predictor coefficients:

$$\hat{x}[n] = -\sum_{k=1}^{p} a[k]x[n-k]$$

by [a] minimizing $\sum_{n=p}^{N}(x[n] - \hat{x}[n])^2$ and [b] minimizing $\mathbb{E}(x[n] - \hat{x}[n])^2$. In the former case, indicate how you would use perturbation theory to calculate approximately the mean and covariance of the prediction coefficients. Also for $p = 3$ draw the lattice realization of the predictor corresponding to case [b].

[2] Let $x[n], n \in \mathbb{Z}$ be a wide sense stationary process with mean $\mu = \mathbb{E}(x[n])$ and covariance

$$C(\tau) = R(\tau) - \mu^2, R(\tau) = \mathbb{E}(x[n+\tau]x[n])$$

Prove the mean ergodic theorem in the L^2 sense:

$$lim_{N\to\infty}(N^{-1}\sum_{n=0}^{N-1} x[n] - \mu)^2 = 0$$

provided that $lim_{|\tau|\to\infty}C(\tau) = 0$.

[3] Let $w[n], n \geq 1$ and $v[n], n \geq 0$ be mutually uncorrelated white noise processes with mean zero and variances σ_w^2 and σ_v^2 respectively. Let

$$s[n] = as[n-1] + w[n], n \geq 1, x[n] = s[n] + v[n], n \geq 0$$

with $|a| < 1$. Assume that $s[0]$ is zero mean and independent of the processes w and v. How will you select $\sigma_s^2 = \mathbb{E}(s[0]^2)$ so that $s[n]$ is WSS. In this case, compute the optimum causal Wiener filter that takes $x[k], k \leq n$ as input and outputs the linear MMSE of $s[n]$ as output. Also write down the Kalman filter equations for this problem. When does the Kalman filter coincide with the Wiener filter ?

[4] Let $\mathbf{X}$ and $\mathbf{Y}$ be two jointly normal random vectors with the following statistics:

$$\mathbb{E}(\mathbf{X}) = \mu_X, \mathbb{E}(\mathbf{Y}) = \mu_Y,$$

$$Cov(\mathbf{X}) = \mathbb{E}((\mathbf{X} - \mu_X)(\mathbf{X} - \mu_X)^T) = \Sigma_{XX},$$

$$Cov(\mathbf{Y}) = \mathbb{E}((\mathbf{Y} - \mu_Y)(\mathbf{Y} - \mu_Y)^T) = \Sigma_{YY},$$

$$Cov(\mathbf{X}, \mathbf{Y}) = \mathbb{E}((\mathbf{X} - \mu_X)(\mathbf{Y} - \mu_Y)^T) = \Sigma_{XY}, \Sigma_{YX} = \Sigma_{XY}^T$$

Show that the MMSE of $\mathbf{X}$ given $\mathbf{Y}$ is given by

$$\hat{X} = \mathbb{E}(\mathbf{X}|\mathbf{Y}) = \mu_X + \Sigma_{XY}\Sigma_{YY}^{-1}(\mathbf{Y} - \mu_Y)$$

Show that

$$Cov(\mathbf{X} - \hat{X}) = \Sigma_{XX} - \Sigma_{XY}\Sigma_{YY}^{-1}\Sigma_{YX}$$

[5] Write short notes on the following.

[a] Innovations process of a WSS process and its application to causal Wiener filtering.

[b] Orthogonal projection operator and its applications to linear signal estimation.

9.1 Problems in Quantum Neural Networks

[1] Explain how you will design a neural network with adaptable weights so that it takes as input a pure quantum state and outputs another pure quantum state.

hint: Let $|\psi_0 >$ be the input state with components $\psi_0(i), i = 1, 2, ..., N$. Assume that the state at the kth layer $|\psi(k) >$ is a pure state for each $k = 1, 2, ..., L$ where $|\psi(L) >$ is the output state. Then, the transition matrix from the kth layer to the $(k+1)^{th}$ layer must be a unitary operator, in general dependent upon the state $|\psi(k) >$. We may thus design this transition matrix $U(k, k+1)$ as

$$U(k, k+1) = exp(iH(k, k+1))$$

where $H(k, k+1)$ is a Hermitian matrix of the form

$$H(k, k+1) = \sum_j W(j, k)A(j, k) + \sum_{j_1, j_2} W(j_1, k)W(j_2, k)A(j_1, j_2, k) + ...$$

$$+ \sum_{j_1, ..., j_n} W(j_1, k)W(j_2, k)...W(j_n, k)A(j_1, ..., j_n, k)$$

where $W(j, k), j = 1, 2, ...$ are the weights of the kth layer and $A(j_1, ..., j_r, k)$ are fixed Hermitian matrices independent of the weights. The output of this QNN is the state

$$|\psi(L) >= U(L-1, L)U(L-2, L-1)...U(1, 2)U(1, 0)|\psi_0 >$$

and the weights $\{W(j,k)\}$ are to be adapted so that $|\psi(L)>$ tracks a given pure state. For implementing this adaptation using the gradient search algorithm, we must differentiate $U(k,k+1)$ w.r.t the weights $W(j,k)$ and to carry out this differentiation, we may use the formula for the differential of the exponential map in Lie group-Lie algebra theory:

$$\frac{\partial}{\partial W(j,k)} U(k,k+1) = \frac{\partial}{\partial W(j,k)} exp(iH(k,k+1))$$

$$= U(k,k+1) \frac{(I - exp(-adH(k,k+1)))}{adH(k,k+1)} \left(\frac{\partial H(k,k+1)}{\partial W(j,k)}\right)$$

[2] Consider a Schrodinger Hamiltonian $H(\theta(t))$:

$$H = H(\theta(t)) = -\nabla^2/2m + V(x,\theta(t))$$

where

$$V(x,\theta(t)) = V_0(x) + \sum_{k=1}^{p} \theta_k(t)V_k(x)$$

The wave function

$$\psi(t,x) = <x|\psi(t)>$$

satisfies

$$i\partial_t|\psi(t)> = H(\theta(t))|\psi(t)>$$

or equivalently,

$$i\partial_t\psi(t,x) = H_0\psi(t,x) + \sum_{k=1}^{p} \theta_k(t)V_k(x)\psi(t,x)$$

where

$$H_0 = -\nabla^2/2m + V_0(x)$$

Adapt the parameters $\theta_k(t)$ so that $|\psi(t,x)|^2 = p(t,x)$ tracks a given probability density $p_0(t,x)$ by approximately minimizing $\int(p_0(t,x) - p(t,x))^2 dx$ using the gradient search algorithm:

$$\theta_k(t+h) = \theta_k(t) - \mu.\frac{\partial}{\partial\theta_k(t)}\int (p_0(t+h,x) - p(t+h,x))^2 dx$$

Now

$$p(t+h,x) = |\psi(t+h,x)|^2$$

$$\psi(t+h,x) = exp(-ih(H_0 + \sum_k \theta_k(t)V_k))\psi(t,x)$$

$$\frac{\partial}{\partial\theta_k} exp(-ih(H_0 + \sum_k \theta_k V_k)) =$$

$$exp(-ih.H)\frac{(I - exp(ih.ad(H)))}{ih.ad(H)}(-ihV_k)$$

$$= h.exp(-ih.H)\sum_{n \geq 0}((ih)^m/(m+1)!)ad(H)^m(V_k)$$

If we make approximations to this formula, then we would use

$$\psi(t+h,x) = \psi(t,x) - ih(H_0\psi(t,x) + \sum_{k=1}^{p}\theta_k(t)V_k(x)\psi(t,x))$$

Note that this approximation defines a non-unitary evolution. Then, we would get

$$\partial\psi(t+h,x)/\partial\theta_k(t) =$$

$$-ihV_k(x)\psi(t,x)$$

and hence

$$\partial p(t+h,x)/\partial\theta_k(t) = 2Re(\bar{\psi}(t+h,x).\partial\psi(t+h,x)/\partial\theta_k(t))$$

$$= 2Re(-ihV_k(x)\bar{\psi}(t+h,x)\psi(t,x)) = 2h.V_k(x).Re(\bar{\psi}(t+h,x)\psi(t,x))$$

9.2 MATLAB simulation exercises in statistical signal processing

[1]

    [a] For $p = 2, 3, 4$ simulate an $AR(p)$ process using the algorithm

$$x[n] = -a[1]x[n-1] - ... - a[p]x[n-p] + w[n]$$

where $\{w[n]\}$ is an iid $N(0, \sigma_w^2)$ sequence with $\sigma_w = 0.1$ and the AR coefficients $a[k], k = 1, 2, ..., p$ are chosen so that the roots of the polynomial

$$A(z) = 1 + a[1]z^{-1} + ... + a[p]z^{-p}$$

fall inside the unit circle. Use the "poly" command to generate the AR polynomial for a specified set of zeroes and use the "root" command to calculate the roots of a polynomial for a specified set of coefficients.

    [b] Write a program to calculate the autocorrelation estimates

$$\hat{R}_{xx}(\tau) = \frac{1}{N-\tau}.\sum_{n=0}^{N-\tau-1}x[n+\tau]x[n]$$

for $\tau = 0, 1, ..., M$ and $\hat{R}_{xx}(-\tau) = \hat{R}_{xx}(\tau), \tau = 1, 2, ..., M$ where $N = 100$ is the number of generated samples of the process and $M << N$, say $M = 20$. Apply the Levinson-Durbin algorithm to calculate the prediction coefficients of the process upto order p. Draw the Lattice filter realization and compare with the lattice filter obtained using the exact statistics of the process, ie, compare the estimated reflection coefficients and the prediction error energies of orders upto p obtained using $\hat{R}_{xx}(\tau)$ with those obtained using $\hat{R}_{xx}(\tau)$. Try to design a perturbation theoretic method for calculating the approximate shift in the reflection coefficients and prediction energies when estimated correlations are used in place of true correlations. Your answer must be expressed in terms of the the the shift

$$\delta R_{xx}(\tau) = \hat{R}_{xx}(\tau) - R_{xx}(\tau)$$

[2] Generate an $ARMA(p,q)$ process with $p = 3, q = 5$ according to the algorithm

$$x[n] + \sum_{k=1}^{p} a[k][n-k] = \sum_{k=0}^{q} b[k]\delta[n-k]$$

where the roots of the polynomial

$$A(z) = 1 + \sum_{k=1}^{p} a[k]z^{-k}$$

all fall inside the unit circle. Here, $\delta[n]$ is the unit impulse function. Estimate $\{a[k]\}$ by minimizing

$$E(\{a[k]\}_{k=1}^{p}) = \sum_{n=max(p,q+1)}^{N} (x[n] + \sum_{k=1}^{p} a[k]x[n-k])^2$$

and then estimate $\{b[k]\}$ using these estimated AR coefficients in the equation

$$b[n] = x[n] + \sum_{k=1}^{p} a[k]x[n-k], n = 0, 1, ..., q$$

[3] Generate an $AR(p)$ process $\{x[n]\}$ as in problem [1]. Generate the desired process,

$$d[n] = c[1]x[n] + c[2]x[n-1] + .. + c[q]x[n-q] + v[n]$$

where $\{v[n]\}$ is iid $N(0, \sigma_v^2)$ independent of $\{w[n]\}$. Take for example $p = 3, q = 2$.

[a] Implement the LMS algorithm for estimating $d[n]$ linearly based on $x[n], x[n-1], ..., x[n-r]$, ie, write

$$\hat{d}[n] = \sum_{k=0}^{r} h_k[n]x[n-k], e[n] - d[n] - \hat{d}[n],$$

$$h_k[n+1] = h_k[n] - \mu \frac{\partial}{\partial h_k[n]} e[n]^2$$

$$= h_k[n] + 2\mu e[n]x[n-k], k = 0, 1, ..., r$$

Plot the error process $e[n]$ and estimate its limiting mean square value using time averages. Now assuming that the vector process

$$\mathbf{X}[n] = [x[n], x[n-1], ..., x[n-r]]^T, n = r, r+1, ...$$

are independent, carry out the standard convergence analysis of the LMS algorithm and determine the limiting mean square error of $e[n]$. Compare this with the estimated one.

[b] Cast the AR filter model and the measurement model for $d[n]$ in standard state variable form and design a Kalman filter for estimating the state, ie, $\xi[n] = [x[n-1], ..., x[n-p]]^T$ from the measured data $d[k], k \leq n$. Also design the causal Wiener filter and compare the asymptotics of the two.

[4] Generate reference signals

$$x_1[n] = cos(\omega n), x_2[n] = sin(\omega n)$$

Generate a signal $y[n] = Acos(\omega n + \phi) + B.cos(\theta n + \psi) + v[n]$ with $\theta \neq \omega$ and $v[n]$ white Gaussian noise. Design an adaptive notch filter based on the LMS algorithm to cancel out the frequency component at ω from $y[n]$ using the reference signals. Specifically, the notch filter is defined by the equations

$$e[n] = y[n] - \hat{y}[n], \hat{y}[n] = h_1[n]x_1[n] + h_2[n]x_2[n],$$

$$h_k[n+1] = h_k[n] - \mu.\frac{\partial}{\partial h_k[n]} e[n]^2 =$$

$$h_k[n] + 2\mu e[n]x_k[n], k = 1, 2$$

Plot the output $e[n]$ of this adaptive notch filter. Also plot its FFT and check that the frequency component at ω has been cancelled out. Solve the above recursive equations for $h_k[n]$ and compare the resulting adaptive notch filter with that obtained using MATLAB simulations.

9.3 Problems in information theory

[1] Let $\phi(x) = x.log(x), x > 0$. ϕ is a convex function since $\phi''(x) = 1/x > 0$. Hence, if p, q are two probability distributions on $E = \{1, 2, ..., N\}$ and $t \in [0, 1]$, then

$$\phi(tp(i) + (1-t)q(i)) \leq t\phi(p(i)) + (1-t)\phi(q(i))$$

Hence, if H denotes the entropy function, then

$$H(tp + (1-t)q) \geq tH(p) + (1-t)H(q)$$

ie, $H(.)$ is a concave function. Further, since $log(x)$ is an increasing function on $\mathbb{R}_+$, it follows that

$$log(tp(i) + (1 - t)q(i)) \geq log(tp(i)),$$

$$log(tp(i) + (1 - t)q(i)) \geq log((1 - t)q(i)),$$

and hence multiplying the first inequality by $tp(i)$, the second inequality by $(1 - t)q(i)$ and adding gives us

$$\phi(tp(i) + (1 - t)q(i)) \geq \phi(tp(i)) + \phi((1 - t)q(i))$$

from which we deduce that

$$H(tp + (1 - t)q) \leq -\sum_i [\phi(tp(i)) + \phi((1 - t)q(i))]$$

$$= H(t, 1 - t) + tH(p) + (1 - t)H(q) \leq log(2) + tH(p) + (1 - t)H(q)$$

We can thus combine the above two inequalities and write,

$$tH(p) + (1 - t)H(q) \leq H(tp + (1 - t)q) \leq log(2) + tH(p) + (1 - t)H(q)$$

Thus, if p_n, q_n are products of p and q respectively on E^n, then

$$tH(p) + (1 - t)H(q) \leq n^{-1}H(tp_n + (1 - t)q_n) \leq log(2)/n + tH(p) + (1 - t)H(q)$$

since $H(p_n) = nH(p), H(q_n) = nH(q)$ and therefore,

$$lim_{n \to \infty} n^{-1}H(tp_n + (1 - t)q_n) = tH(p) + (1 - t)H(q)$$

Entropy rate of a stationary dynamical system: Let $(\Omega, \mathcal{F})$ be a measurable space and let P be a probability measure on this space. Let $T : (\Omega, \mathcal{F}) \to (\Omega, \mathcal{F})$ be a measurable map that preserves P, ie $P(T^{-1}E) = P(E)\forall E \in \mathcal{F}$. We write this for short as, $PoT^{-1} = P$. Let ξ be a finite measurable partition of $(\Omega, \mathcal{F})$, ie, $\xi = \{E_1, ..., E_N\}$ for some finite positive integer N where $E_i \in \mathcal{F}$, $E_i \cap E_j = \phi, i \neq j$ and $\Omega = \bigcup_{i=1}^N E_i$. Given two such partitions ξ, η (by partition, we mean measurable partition, ie, all its elements are members of $\mathcal{F}$), we define

$$\xi \vee \eta = \{E \cap F : E \in \xi, F \in \eta\}$$

Then $\xi \vee \eta$ is also a finite partition of Ω. Given two partitions ξ, η of Ω, we write $\xi > \eta$, if every element of η is a union of elements of ξ. It immediately follows that for any two partitions ξ, η, we have that $\xi \vee \eta > \xi, \eta$.

9.4 Problems in quantum neural networks

[1] Let $H(\theta)$ be a Hamiltonian operator depending upon a vector parameter θ. Assume that the eigenvalue problem

$$H(\theta)|u_n(\theta) >= E_n(\theta)|u_n(\theta) >, n = 0, 1, 2, ...$$

has been solved. Now suppose $\theta = \theta(t)$ is made to vary with time. Then, we have

$$H(\theta(t))|u_n(\theta(t)) >= E_n(\theta(t))|u_n(\theta(t)) >, n = 0, 1, 2, ...$$

We wish to solve the time dependent Schrodinger equation

$$i\frac{d}{dt}|\psi(t) >= H((\theta(t))|\psi(t) >$$

Expand the solution as

$$|\psi(t) >= \sum_{n\geq 0} c(n, t)|u_n(\theta(t)) >$$

The problem is to derive an system of coupled linear differential equations for $c(n, t), n \geq 0$.

$$i\partial_t|\psi(t) >= H(\theta(t))|\psi(t) >$$

$$H(\theta(t))|u_n(\theta(t)) >= E_n(\theta(t))|u_n(\theta(t)) >$$

$$|\psi(t) >= \sum_n c(n, t)|u_n(\theta(t)) >$$

$$i\partial_t|\psi(t) >= \sum_n [i(\partial_t c(n, t))|u_n(\theta(t)) > +ic(n, t)\partial_t|u_n(\theta(t)) >]$$

$$\partial_t|u_n(\theta(t)) >= \sum_k \theta'_k(t)\partial_{\theta_k}|u_n(\theta(t)) >$$

$$< u_m(\theta(t))|\partial_t|u_n(\theta(t)) >= \sum_k \theta'_k(t) < u_m(\theta(t))|\partial_{\theta_k}|u_n(\theta(t)) >$$

Define

$$F(m, n, k, \theta) =< u_m(\theta)|\partial_{\theta_k}|u_n(\theta) >$$

Suppose that $|u_n(\theta) >$ are real functions of θ in the position representation. Then, $F(m, n, k, \theta)$ are also real functions. In the general case, the equation

$$< u_m(\theta)|u_n(\theta) >= \delta(m, n)$$

implies on differentiating that

$$F(m, n, k, \theta) + \bar{F}(n, m, k, \theta) = 0$$

In particular, if $F(n, n, k, \theta)$ is real, it follows that

$$F(n, k, k, \theta) = 0$$

In the general case, we get Schrodinger's equation as

$$i\partial_t c(n, t) + i \sum_m c(m, t) < u_n(\theta(t))|\partial_t|u_m(\theta(t)) >= E_n(\theta(t))c(n, t)$$

or equivalently,

$$i\partial_t c(n, t) + i \sum_{m,k} c(m, t)F(n, m, k, \theta(t))\theta_k'(t)c(m, t) = E_n(\theta(t))c(n, t) - - - (1)$$

Neural network design: Let $p_0(t, q)$ be the density to be tracked by $p(t, q) = |\psi(t, q)|^2$ where $\psi(t, q) =< q|\psi(t) >$ is the position space representation of $|\psi(t) >$. To achieve this tracking, we can either adaptively minimize $\int (p_0(t, q) - |\psi(t, q)|^2)^2 dq$, or else adaptively minimize $\sum_n |c_0(n, t) - c(n, t)|^2$ where

$$c_0(n, t) = \int \sqrt{p_0(t, q)} \bar{u}_n(\theta(t), q)dq =< u_n(\theta(t))|\sqrt{p_0(t, .)} >$$

9.5   Quantum Gaussian states and their transformations

$a = (a_1, ..., a_n), , a^* = (a_1^*, ..., a_n^*)$ form a canonical set of annihilation and creation operators:

$$[a_i, a_j] = 0, [a_i, a_j^*] = \delta(i, j)$$

The Weyl operator is defined as

$$W(z) = exp(z^*a - z^T a^*), z \in \mathbb{C}^n$$

Note that

$$W(z) = exp(-|z|^2/2).exp(-z^T a^*).exp(z^*a)$$

which implies

$$< e(v)|W(z)|e(u) >= exp(-|z|^2/2+ < z, u > - < v, z >)$$

A state ρ will be called zero mean Gaussian if

$$Tr(\rho.W(z)) = exp(R(z)^T SR(z)/2)$$

where

$$z = x + iy, R(z) = [x^T, y^T]^T$$

and S is a symmetric complex matrix. Note that

$$W(z)|e(u)> = exp(-|z|^2/2)exp(-z^T a^*)exp(<z,u>)|e(u)>$$

$$= exp(-|z|^2/2+ <z,u>)|e(u-z)>$$

$$W(-z)W(z)|e(u)> = exp(-|z|^2/2+ <z,u>)W(-z)|e(u-z)> =$$

$$exp(-|z|^2/2+ <z,u>).exp(-|z|^2/2- <z,u-z>) = 1$$

so $W(z)$ is a unitary operator with inverse $W(-z)$:

$$W(z)^* = exp(z^T a^* - z^* a) = W(-z) = W(z)^{-1}$$

Also,

$$W(v)W(z)|e(u)> = exp(-|z|^2/2+ <z,u>)W(v)|e(u-z)>$$

$$= exp(-|z|^2/2+ <z,u>).exp(-|v|^2/2+ <v,u-z>)|e(u-z-v)>$$

$$= exp(-|z|^2/2 - |v|^2/2- <v,z> + <z+v,u>)|e(u-z-v)>$$

$$= exp(-|z+v|^2+ <z+v,u> -iIm(<v,z>))|e(u-z-v)>$$

$$= exp(-iIm(<v,z>))W(z+v)|e(u)>$$

Therefore, $z \to W(z)$ is a projective unitary representation of $\mathbb{C}^n$ in the Boson Fock space $\Gamma_s(\mathbb{C}^n) = L^2(\mathbb{R}^n)$:

$$W(v)W(z) = exp(-iIm(<v,z>))W(v+z)$$

Now let ρ be a zero mean Gaussian state. Then

$$\hat{\rho}(z) = Tr(\rho.W(z)) = exp(-R(z)^T SR(z)/2)$$

now for p points $z_1,...,z_p \in \mathbb{C}^n$, we have that

$$((Tr(\rho W(z_i)W(z_j)^*))_{1\le i,j\le n}$$

is a positive definite matrix since ρ is a state and hence positive definite. But

$$W(z_i)W(z_j)^* = W(z_i)W(-z_j) = exp(iIm(<z_i,z_j>))W(z_i - z_j)$$

and hence

$$((\hat{\rho}(z_i - z_j)exp(iIm(<z_i,z_j>))))_{1\le i,j\le p}$$

is a positive definite matrix. Now if $\xi = R(z), \eta = R(u)$ with $z = x + iy, u = a + ib$, then $\xi = [x^T, y^T]^T, \eta = [a^T, b^T]^T$ and $J\eta = [b^T, -a^T]^T$ so that

$$< R(z), JR(u) > = < \xi, J\eta > = x^T b - y^T a = Im(<z,u>)$$

So

$$Im(<z_i, z_j>) = < R(z_i), JR(z_j) >$$

Further,

$$0 = < R(z_i), JR(z_i) > = < R(z_j), JR(z_j) >$$

We require that S be such that S is symmetric and

$$((exp((-1/2)(R(z_i - z_j)^T SR(z_i - z_j) + i < R(z_i), JR(z_j) >))) - - - (1)$$

is positive definite. A necessary and sufficient condition for this is that

$$S + iJ \geq 0$$

Remark: For (1) to be positive definite, it is necessary and sufficient that the matrix $A = (((-1/2)(\xi_i - \xi_j)^T S(\xi_i - \xi_j) + i\xi^T J\xi_j))$ be conditionally positive definite for all real vectors $\xi_i, i = 1, 2, ...p$ of size $2n \times 1$. This means that if $\sum c_i = 0$, then

$$\sum_{i,j} \bar{c}_i c_j A(i, j) \geq 0$$

Equivalently, if $\sum_i c_i = 0$, then

$$\sum_{i,j} \bar{c}_i c_j (\xi_i^T S\xi_j + i\xi_i^T J\xi_j) \geq 0$$

or equivalently,

$$S + iJ \geq 0$$

Assume that this condition has been satisfied. Consider now the following transformation acting on zero mean Gaussian states. The state ρ gets transformed to $T^*(\rho)$ where T^* is the dual of the linear operator T acting on bounded operators in $\Gamma_s(\mathbb{C}^n)$ defined by its action on $W(z)$:

$$T(W(z)) = W(R^{-1}AR(z)).exp((1/2)(R(z)^T BR(z)))$$

where A is a real matrix of size $2n \times 2n$ and B is a real symmetric $2n \times 2n$ matrix. Under this transformation, it is easily seen that if (A, B) satisfies a certain condition, then $T^*(\rho)$ is a zero mean Gaussian state whenever ρ is. Consider a one parameter family $(A_t, B_t), t \in \mathbb{R}$ of matrices such that if

$$T_t(W(z)) = W(R^{-1}A_t R(z))exp((1/2)R(z)^T B_t R(z))$$

then

$$T_{t+s} = T_t T_s \forall t, s \in \mathbb{R}$$

ie T_t is a one parameter group of transformations on the Banach space of bounded operators on the Boson Fock space $\Gamma_s(\mathbb{C}^n)$ such that its dual $T_t^*, t \in \mathbb{R}$ is a one parameter group of transformations on the convex set of states in $\Gamma_s(\mathbb{C}^n)$ that preserves Gaussian states. Then we have

$$W(R^{-1}A_{t+s}R(z))exp((1/2)R(z)^T B_{t+s}R(z)) =$$

$$T_t(W(R^{-1}A_sR(z))exp((1/2)R(z)^T B_s R(z)))$$

$$= W(R^{-1}A_t A_s R(z))exp((1/2)R(z)^T B_s R(z) + (1/2)R(z)A_s^T B_t A_s R(z))$$

This is to be supplemented with the condition $T_0 = I$. This is satisfied iff

$$A_{t+s} = A_t A_s, B_{t+s} = B_s + A_s^T B_t A_s \forall t, s \in \mathbb{R}$$

This is satisfied iff

$$dA_s/ds = LA_s, L = (dA_t/dt)|_{t=0}$$

and

$$dB_t/dt = K + L^T B_t + B_t L, K = (dB_s/ds)|_{s=0}, A_0 = I, B_0 = 0$$

which is satisfied iff

$$B_t = \int_0^t exp((t-s)L^T)K.exp((t-s)L)ds = \int_0^t exp(sL^T)K.exp(sL)ds,$$

$$A_t = exp(tL)$$

Thus, we get a parametrization of T_t in terms of the operators (L, K). Now let ρ be a zero mean Gaussian state with covariance matrix S, ie, S is a $2n \times 2n$ complex matrix with

$$S + iJ \geq 0$$

and

$$Tr(\rho.W(z)) = exp((1/2)R(z)^T S R(z))$$

Then,

$$Tr(T_t^*(\rho)W(z)) = Tr(\rho.T_t(W(z)))$$

$$= Tr(\rho.W(R^{-1}A_t R(z))).exp((1/2)R(z)^T B_t R(z))$$

$$= \hat{\rho}(R^{-1}A_t R(z)).exp((1/2)R(z)^T B_t R(z))$$

$$= exp((1/2)R(z)^T(B_t + A_t^T S A_t)R(z))$$

For this to be a valid Gaussian state, an additional condition on (A_t, B_t) must be imposed, namely,

$$A_t^T S A_t + B_t + iJ \geq 0 \forall t --- (a)$$

Problem: How to construct Lindblad operators and a harmonic oscillator Hamiltonian so that condition (a) is satisfied.

Remarks: let ρ be any state with quantum Fourier transform $\phi(z) = \hat{\rho}(z)$. Let $\phi_t(z)$ be the quantum Fourier transform of $\rho_t = T_t^*(\rho)$. Thus,

$$\phi_t(z) = \phi(R^{-1}A_t R(z))exp(1/2)R(z)^T B_t R(z))$$

where

$$A_t = exp(tL), B_t = \int_0^t exp(sL^T)K.exp(sL)ds$$

Then,

$$\frac{d\phi_t(z)}{dt}|_{t=0} = (R^{-1}LR(z))^T$$

More precisely, ϕ and ϕ_t are functions of $z, \bar{z}$ or equivalently of $R(z) = \xi = [x^T, y^T]^T$. So we should write

$$\phi_t(\xi) = \phi_t(R(z)) = \phi_t(\xi)$$

Then, with this notation, we have

$$\phi_t(\xi) = \phi(A_t\xi)exp((1/2)\xi^T B_t\xi)$$

and we find for the action of the generator of T_t^* on $\phi(\xi)$ the formula

$$\mathcal{L}\phi(\xi) = \frac{d}{dt}\phi_t(\xi)|_{t=0} =$$

$$(L\xi)^T \nabla_\xi \phi(\xi) + (1/2)(\xi^T K\xi)\phi(\xi)$$

We write $\phi_\rho(\xi) = Tr(\rho W(z)) = \phi(\xi)$ to show the explicit dependence of ϕ on ρ. Then,

$$\xi.\phi(\xi) = (x\phi(\xi), y\phi(\xi))$$

with

$$x\phi(\xi) = Tr(\rho.xW(z)), y\phi(\xi) = Tr(\rho.yW(z))$$

so that

$$\xi.\phi(\xi) = Tr(\rho.\xi.W(z))$$

Now,

$$W(z)|e(u) >= exp(-|z|^2/2+ < z, u >)|e(u - z) >$$

so that

$$aW(z)|e(u) >= (u - z)exp(-|z|^2/2+ < z, u >)|e(u - z) >,$$

$$W(z)a|e(u) >= u.exp(-|z|^2/2+ < z, u >)|e(u - z) >$$

so

$$[W(z), a] = zW(z)$$

or equivalently,

$$W(z)aW(z)^* - a = z$$

and we get on writing

$$a = (q + ip)/\sqrt{2}, a^* = (q - ip)/\sqrt{2}$$

that

$$(W(z)qW(z)^* - q) + i(W(z)pW(z)^* - p) = (x + iy)\sqrt{2}$$

which yields on equating Hermitian and skew-Hermitian parts on both sides,

$$W(z)qW(z)^* - q = x\sqrt{2}, W(z)pW(z)^* - p = y\sqrt{2}$$

or equivalently,

$$[W(z), q] = \sqrt{2}xW(z), [W(z), p] = \sqrt{2}yW(z)$$

and hence

$$x\phi_\rho(\xi) = Tr(\rho.xW(z)) = 2^{-1/2}Tr(\rho.[W(z), q])$$
$$= 2^{-1/2}Tr([q, \rho]W(z)) = 2^{-1/2}\phi_{[q,\rho]}(\xi)$$

Likewise,

$$y\phi_\rho(\xi) = 2^{-1/2}\phi_{[p,\rho]}(\xi)$$

Further,

$$\nabla_x\phi(\xi) = Tr(\rho.\nabla_xW(z))$$

Now,

$$W(z) = exp(-|z|/2)exp(-z^Ta^*)exp(z^*a), z = x + iy$$

so

$$\nabla_xW(z) = -xW(z) - a^*W(z) + W(z)a = -2^{-1/2}[W(z), q] + W(z)a - a^*W(z)$$

$$\nabla_yW(z) = -yW(z) - ia^*W(z) - iW(z)a = -2^{-1/2}[W(z), p] - i(W(z)a + a^*W(z))$$

So

$$\nabla_x\phi(\xi) = Tr(\rho.\nabla_xW(z)) = -2^{-1/2}\phi_{[q,\rho]}(\xi) + \phi_{a\rho}(\xi) - \phi_{\rho.a^*}(\xi)$$
$$\nabla_y\phi(\xi) = Tr(\rho.\nabla_yW(z)) = -2^{-1/2}\phi_{[p,\rho]}(\xi) - i\phi_{a\rho}(\xi) - i\phi_{\rho.a^*}(\xi)$$

Chapter 10

Lecture Plan for Information Theory, Sanov's Theorem, Quantum Hypothesis Testing and State Transmission, Quantum Entanglement, Quantum Security

10.1 Lecture Plan

[1] Shannon's construction of the entropy function from basic postulates.

    [2] The meaning of information and its relationship to probability.

    [3] Concavity of the entropy function.

    [4] Shannon's noiseless coding theorem or the data compression theorem.

    [a] Chebyshev's inequality and the weak law of large numbers.

    [b] (ϵ, n)-typical sequences.

    [c] Shannon's noiseless coding theorem for discrete memoryless sources.

    [d] Ergodic sources and the Shannon-McMillan-Breiman theorem:Proof based on the Martingale convergence theorem.

    [e] Entropy rate of a stationary stochastic process.

    [f] Noiseless coding theorem for ergodic sources.

    [g] Converse of the noiseless coding theorem for iid and ergodic sources.

    [5] No prefix codes, uniquely decipherable codes and inequalities for the

average length of a uniquely decipherable code.

[6] Proof that if Kraft's inequality is satisfied, a no prefix code can be constructed, if a code is uniquely decipherable, then Kraft's inequality is satisfied and that no prefix codes are uniquely decipherable.

[7] Minimum average length of a code and construction of the optimal Huffman code.

[8] Basic large deviation theory required for proving the rate distortion theorem.

[9] Proof of the rate distortion theorem and its relationship to the generalization of Shannon's noiseless compression theorem with maximum allowable distortion.

[a] Proof based on the large deviation principle with random coding argument.

[b] Proof based on a random coding argument, distortion typical sequence pairs without the use of large deviations.

[10] Noisy channels, conditional entropy and mutual information.

[11] Shannon's noisy coding theorem and information capacity of a discrete memoryless channel.

[a] Jointly typical sequences.

[b] Proof of the noisy coding theorem based on Feinstein-Khintchine fundamental lemma.

[c] Proof of the converse of the noisy coding theorem.

[d] Joint input-output ergodicity for stationary channels having finite memory and the m-independence property when the input is ergodic.

[e] Shannon's proof of the noisy coding theorem based on joint typicality and a random coding argument.

[12] Rate distortion theorem as a sphere covering problem and the noisy coding theorem as a sphere packing problem.

[13] Relative entropy and mutual information as a distance measure between two probability distributions.

[14] Joint convexity of the relative entropy.

[15] Kolmogorov-Sinai entropy of a stationary dynamical system and its properties.

[16] Sanov's theorem on relative entropy as the rate function for empirical distributions of iid random variables.

[17] Capacity of Gaussian channels.

[18] An introduction to network information theory.

[a] Capacity region of multiple access channels.

[19] An introduction to quantum information theory.

[a] Entropy of a quantum state.

[b] Quantum hypothesis testing between two states and quantum Stein's lemma.

[c] Concavity of quantum entropy.

[d] Joint convexity of quantum relative entropy, proof based on application of Lieb's concavity theorem to quantum relative Renyi entropy.

[e] Quantum data compression using Schumacher's noiseless coding theorem,

[f] Quantum relative Renyi entropy and its joint convexity based on operator convex functions.

[g] Cq noisy coding theorem and its converse. Proofs based on entropy typical projections and random coding arguments.

10.2 A problem in information theory

Let $p(i), i = 1, 2, ..., N$ be a probability distribution on N symbols, say on $E = \{1, 2, ..., N\}$. Assume that $p(1) \geq p(2) \geq ... \geq p(N)$. Let $p_n(x), x \in E^n$ denote the product probability distribution on E^n, i.e, if $x = (x_1, ..., x_n)$ with $x_i \in E, i = 1, 2, ..., N$, then $p_n(x) = p(x_1)...p(x_n)$. Let $p(x)^{\downarrow}$ denote the arrangement of the numbers $p(x), x \in E^n$ in decreasing order. Let $|F|$ denote the size of a set F. Then,

$$|\{x : p(x) > exp(b)\}| \leq \sum_x (p(x)/e^b)^{1-s} = exp(\psi(s) - b(1-s))$$

where

$$\psi(s) = log \sum_x p(x)^{1-s}$$

and $0 \leq s \leq 1$. Thus writing

$$\psi(s) - (1-s)b = R$$

we get that

$$b = b(s) = (\psi(s) - R)/(1-s)$$

and we get in this case,

$$|\{x : p(x) > exp(b(s))\}| \leq exp(R)$$

It follows then that if

$$P(p, e^R) = \sum_{x \leq exp(R)} p(x)$$

then

$$p(x) \leq exp(b(s)), x > exp(R)$$

and hence

$$P(p, e^R)^c = 1 - P(p, e^R) = \sum_{x > exp(R)} p(x) \leq \sum_{\{x:p(x) \leq exp(b(s))\}} p(x)$$

$$\leq \sum_x p(x)(exp(b(s))/p(x))^s = \sum_x p(x)^{1-s} exp(sb(s))$$

$$= exp(sb(s) + \psi(s))$$

Now,

$$sb(s) + \psi(s) = s(\psi(s) - R)/(1 - s) + \psi(s) = (\psi(s) - sR)/(1 - s)$$

and hence,

$$P(p, R)^c \leq exp(min_{0 \leq s \leq 1}(\psi(s) - sR)/(1 - s)) - - - (1)$$

We now derive an upper bound on $P(p, e^R)$. Let A be any set with $|A| = e^R$. Then

$$A = (A \cap \{x : p(x) > a\}) \cup (A \cap \{x : p(x) \leq a\})$$

so that

$$p(A) \leq a|A| + p\{x : p(x) > a\}$$

In particular, we get with $a = e^b, b = b(s) = (\psi(s) - sR)/(1 - s)$ and $s < 0$,

$$p(p, e^R) = \sum_{x=1}^{e^R} p(x) \leq e^{R+b} + \sum_x p(x)(p(x)/e^b)^{-s}$$

$$= exp((\psi(s) - sR)/(1 - s)) + exp(\psi(s) + sb) = 2exp((\psi(s) - sR)/(1 - s))$$

where we have used

$$R + b = R + (\psi(s) - R)/(1 - s) = (\psi(s) - sR)/(1 - s), \psi(s) + sb = \psi(s) + s(\psi(s) - R)/(1 - s)$$

$$= (\psi(s) - sR)/(1 - s)$$

Since $s < 0$ is arbitary, it follows that

$$p(p, e^R) \leq 2.exp(min_{s<0}(\psi(s) - sR)/(1 - s)) - - - (2)$$

In particular, applying (1) and (2) to p_n in place of p and nR in place of R gives us

$$p(p_n, e^{nR})^c \leq exp(n.min_{0 \leq s < 1}(\psi(s) - sR)/1 - s))),$$

$$p(p_n, e^{nR}) \leq 2.exp(n.min_{s<0}(\psi(s) - sR)/(1 - s))$$

where

$$\psi(s) = log \sum_x p(x)^{1-s}$$

Now observe that

$$lim_{s \to 0}\psi(s)/s = -\sum_x p(x).log(p(x)) = H(p)$$

and therefore, it easily follows that if $R > H(p)$, then

$$limsup_n(n^{-1}.log(p(p_n, e^{nR})^c)) < 0$$

and in particular,

$$lim_{n \to \infty}p(p_n, e^{nR})^c = 0$$

or equivalently,
$$lim_n p(n, e^{nR}) = 1$$

and if $R < H(p)$, then

$$limsup_n n^{-1}.log(p(p_n, e^{nR})) < 0$$

and in particular,
$$lim_{n \to \infty} p(p_n, e^{nR}) = 0$$

Remark: This result can be looked upon as a version of Shannon's noiseless coding theorem based on typical sequences, ie, the set of typical sequences consists of approximately $exp(nH(p))$ elements and this set occupies almost all the probability mass in the set of all the $exp(n.logN) = N^n$ sequences. Note that, $H(p) < N$ except when p is the uniform distribution and hence for sequences of large length n, $exp(nH(p)) << N^n$.

10.3 Types and Sanov's theorem

The same situation as above is being considered. Let $u \in E^n$ where $E = \{1, 2, ..., N\}$ and let p_u denote the type probability distribution corresponding to u, ie,

$$p_u(x) = N(x|u)/n, x \in E$$

where $N(x|u)$ is the number of times that x occurs in u. Thus a type (probability distribution) is a probability distribution on E of the form

$$p(x) = m(x)/n, x \in E$$

where $m(x), x \in E$ are all non-negative integers. It is clear that the total number of types is $\le (n+1)^{N-1}$ since each element in E can occur either 0 times, or once, or twice,....,or n times and that the number of times that the last element N of E occurs is determined by the number of times that the other elements have occurred since the sum of the number of occurrences of all the elements is n. Let p be a type and let $T(n, p)$ denote the set of all sequences of type p. Then,

$$|T(n, p)| = \frac{n!}{\Pi_{x \in E}(np(x))!}$$

Further if $u \in T(n, p)$, and p_n denotes the n-fold product distribution of p, then

$$p_n(u) = \Pi_{x \in E} p(x)^{np(x)} = exp(n \sum_x p(x).log(p(x))) = exp(-nH(p))$$

Let now p, q be two types. Then, for $u \in T(n, q)$,

$$p_n(u) = \Pi_x p(x)^{nq(x)} = exp(\sum_x nq(x)log(p(x))) = exp(-n(H(q) + D(q|p))$$

Further,
$$|T(n,p)|/|T(n,q)| = \Pi_x[(nq(x))!/(np(x))!]$$

Now,
$$n!/m! \leq n^{n-m}$$

proof: Suppose $n \geq m$. Then, lhs is $n(n-1)..(n-m+1) \leq n^{n-m}$. Suppose $n < m$. Then reciprocal of lhs is $m(m-1)...(m-n+1) \geq n^{m-n}$. The proof is complete.

Now, using this result,

$$|T(n,q)|/|T(n,p)| = \Pi_x((np(x))!/(nq(x))!)$$

$$\leq \Pi_x(np(x))^{n(p(x)-q(x))} = exp(\sum_x n(p(x)-q(x))log(np(x)))$$

$$= exp(n(\sum_x(p(x)log(p(x))-q(x)log(p(x))))) = exp(-n(H(p)-H(q)-D(q|p)))$$

Therefore, for any $u \in T(n,p), v \in T(n,q)$,

$$p_n(T(n,q))/p_n(T(n,p)) = p_n(v).|T(n,p)|/p_n(u)|T(n,q)|$$

$$= exp(-n(H(q)+D(q|p)-H(p)))|T(n,q)|/|T(n,p)| \leq 1$$

Thus, we obtain

Theorem: For any two types p,q, we have

$$p_n(T(n,p)) \geq p_n(T(n,q))$$

Now, $u \in T(n,q)$ implies

$$p_n(u) = exp(-n(H(q)+D(q|p)))$$

and summing this equation over all $u \in T(n,q)$ results in

$$p_n(T(n,q)) = exp(-n(H(q)+D(q|p)))|T(n,q)|$$

Since this is a probability, it cannot exceed unity and hence we get

$$|T(n,q)| \leq exp(n(H(q)+D(q|p)))$$

This inequality is true for any two types p,q. In particular, setting $p=q$ gives

$$|T(n,q)| \leq exp(nH(q))$$

or equivalently, replacing q by p,

$$|T(n,p)| \leq exp(nH(p))$$

Now we have seen that for any two types p,q,

$$exp(-nH(p))|T(n,p)| = p_n(T(n,p)) \geq p_n(T(n,q))$$

Since if q, q' are any two different types, $T(n, q)$ and $T(n, q')$ are disjoint, it follows then by summing over all types q and recalling that the number of types cannot exceed $(n + 1)^{N-1}$, that

$$exp(-nH(p))|T(n, p)|(n + 1)^{N-1} \geq \sum_q p_n(T(n, q)) = 1$$

or equivalently,

$$|T(n, p)| \geq exp(nH(p))/(n + 1)^{N-1}$$

Thus, we have proved
    Theorem: For any type p,

$$exp(nH(p))/(n + 1)^{N-1} \leq |T(n, p)| \leq exp(nH(p))$$

In particular,

$$lim_{n \to \infty} \frac{log(|T(n, p)|)}{n} = H(p)$$

Combining this equation with the equation

$$p_n(T(n, q)) = exp(-n(H(q) + D(q|p))|T(n, q)|$$

derived above, gives us
    Theorem (Sanov): For any two types p, q,

$$exp(-nD(q|p))/(n + 1)^{N-1} \leq p_n(T(n, q)) \leq exp(-nD(q|p))$$

and in particular,

$$lim_{n \to \infty} n^{-1}.log(p_n(T(n, q)) = -D(q|p)$$

Intuitively, this means that if the relative entropy distance $D(q|p)$ between q and p is large, then the probability that under p, the empirical distribution of the sequence will be q is less likely and vice-versa.

10.4 Quantum Stein's theorem

We wish to discriminate between two states ρ, σ using a sequence of iid measurements, ie, measurements taken when the states become $\rho^{\otimes n}$ and $\sigma^{\otimes n}, n = 1, 2, ...$ Let $0 \leq T_n \leq 1$ be the detection operators. If the state is $\rho^{\otimes n}$, then the probability of a wrong decision is $P_1(n) = Tr(\rho^{\otimes n}(1 - T_n))$ while if the state is $\sigma^{\otimes n}$, the probability of a wrong decision is $P_2(n) = Tr(\sigma^{\otimes n}T_n)$. The theorem that we prove is the following: If the measurement operators $\{T_n\}$ are such that $P_2(n) \to 0$, then $P_1(n)$ cannot go to zero at a rate faster than $exp(-nD(\sigma|\rho))$, ie,

$$liminf_{n \to \infty} n^{-1}.log(P_1(n)) \geq -D(\sigma|\rho)$$

and conversely, for any $\delta > 0$, there there exists a sequence $\{T_n\}$ of detection operators such that $P_2(n) \to 0$ and simultaneously

$$limsup_n n^{-1}.log(P_1(n)) \leq -D(\sigma|\rho) + \delta$$

To prove this, we first use the monotonicity of the Renyi entropy to deduce that for any detection operator T_n,

$$(Tr(\rho^{\otimes n}(1 - T_n)))^s Tr(\sigma^{\otimes n}(1 - T_n))^{1-s}$$

$$\leq Tr(\rho^{\otimes ns})\sigma^{\otimes n(1-s)})$$

for any $s \leq 0$. Therefore, if

$$Tr(\sigma^{\otimes n} T_n) \to 0$$

then

$$Tr(\sigma^{\otimes n}(1 - T_n)) \to 1$$

and we get from the above on taking logarithms,

$$limsup n^{-1}.log(Tr(\rho^{\otimes n}(1 - T_n)) \geq log(Tr(\rho^s \sigma^{1-s}))/s$$

or equivalently writing

$$\alpha = limsup n^{-1}.log(P_1(n))$$

we get

$$\alpha \geq s^{-1} Tr(\rho^s \sigma^{1-s}) \forall s \leq 0$$

Taking $lim s \uparrow 0$ in this expression gives us

$$\alpha \geq -D(\sigma|\rho) = -Tr(\sigma(log(\sigma) - log(\rho)))$$

This proves the first part. For the second part, we choose T_n in accordance with the optimal Neyman-Pearson test. Specifically, choose

$$T_n = \{exp(nR)\rho^{\otimes n} \geq \sigma^{\otimes n}\}$$

where the real number R will be selected appropriately. We then find the following upper bound for the probability of false alarm:

$$Tr(\sigma^{\otimes n} T_n) \leq exp(nsR)Tr(\rho^{\otimes n})^s(\sigma^{\otimes n})^{1-s})$$

$$= exp(ns(R + s^{-1}.log(Tr(\rho^s \sigma^{1-s})))) --- (a)$$

for any $s > 0$ and for the probability of miss,

$$Tr(\rho^{\otimes n}(1 - T_n)) \leq Tr(\rho^{\otimes n})^{1-s}(\sigma^{\otimes n})^s)exp(-nsR))$$

$$= exp(-ns(R - s^{-1}log(Tr(\rho^{1-s}\sigma^s)))) --- (b)$$

for any $s > 0$. Now let $\delta > 0$ be arbitrary. Choose

$$R = D(\sigma|\rho) - \delta$$

Then since

$$lim_{s\downarrow 0}s^{-1}.log(Tr(\rho^s\sigma^{1-s})) = -D(\sigma|\rho)$$

it follows that for $s = s_0$ positive and sufficiently small that

$$s_0^{-1}log(Tr(\rho^s\sigma^{1-s})) < -D(\sigma|\rho) + \delta/2$$

and then we get from (a)

$$Tr(\sigma^{\otimes n}T_n) \leq exp(-ns_0\delta/2) \to 0$$

which says that for this sequence of tests, the probability of false alarm converges to zero, on the one hand while on the other, we find with $s = 1$ in (b) that

$$Tr(\rho^{\otimes n}(1 - T_n)) \leq exp(-n(D(\sigma|\rho) - \delta))$$

which shows that the probability of miss for the same sequence of tests converges to zero at a rate faster than $D(\sigma|\rho) - \delta$.

10.5   Problems in statistical image processing

[1] Explain the meaning of orthogonal transforms acting on vectors (ie,one dimensional signals) and matrices (ie two dimensional image fields). Prove that given two one dimensional orthogonal transforms, their tensor product defines an orthogonal transform acting on two dimensional image fields.

[2] Prove that given K one dimensional orthogonal transforms defined by matrices $T_1, ..., T_K$ as acting on one dimensional signals of perhaps different sizes, their tensor product $T_1 \otimes ... \otimes T_K$ defines an orthogonal transform acting on K-dimensional signals.

[3] Define the Hadamard transform

$$H = 2^{-1/2} \begin{pmatrix} 1 & 1 \\ 1 & -1 \end{pmatrix}$$

acting on a signal having just two samples. Show that H is an orthogonal transform. Consider $H_n = H^{\otimes n}$. Show that H_n is a $2^n \times 2^n$ orthogonal matrix acting on a one dimensional signal having 2^n components. Write down explicitly the matrices H_n for $n = 1, 2, 3$. Show that $H_n \otimes H_n$ is an orthogonal transform acting on a $2^n \times 2^n$ dimensional image field.

[4] Consider the one dimensional vibrating string $u(t, x)$ satisfying the one dimensional wave equation with forcing $f(t, x)$.

$$\frac{\partial^2 u(t, x)}{\partial t^2} - c^2 \frac{\partial^2 u(t, x)}{\partial x^2} = f(t, x), 0 \leq x \leq L$$

with boundary conditions $u(t, 0) = u(t, L) = 0$.

$u(t, x)$ can be regarded as a two dimensional random field provided that $f(t, x)$ is a random field. Solve this equation explicitly for $u(t, x)$ assuming zero initial conditions (ie, we are calculating only the forced response), by first solving for the Green's function

$$\frac{\partial^2 G(t, x|t', x')}{\partial t^2} - c^2 \frac{\partial^2 G(t, x|t', x')}{\partial x^2} = \delta(t - t')\delta(x - x') - - - (1)$$

with boundary conditions

$$G(t, 0|t'.x') = G(t, L|t', x') = 0$$

Show that this equation can be solved by using

$$\delta(x - x') = (2/L) \sum_{n \geq 1} sin(n\pi x/L).sin(n\pi x'/L), 0 \leq x, x' \leq L$$

and setting

$$G(t, x|t', x') = \sum_n G_n(t|t', x')sin(n\pi x/L)$$

giving thereby

$$\partial_t^2 G_n(t|t',x') + (n\pi c/L)^2 G_n(t|t',x') = (2/L)sin(n\pi x'/L)\delta(t-t') ---- (2)$$

Deduce from (2) the boundary conditions,

$$G_n(t-0|t',x') = G_n(t+0|t',x')$$

$$= (2/L)sin(n\pi x'/L), \partial_t G_n(t-0|t',x') = \partial_t G_n(t+0|t',x') = 0,$$

and hence solve for G_n and thus evaluate G_n and hence G. Show that the solution to the wave equation with random forcing can be expressed as

$$u(t,x) = \int G(t,x|t',x')f(t',x')dt'dx'$$

Calculate the correlation function of the Guassian random field $u(t,x)$ assuming that $f(t,x)$ is a zero mean Gaussian random field. Assume that the parameters c, L of the string ($c = \sqrt{T/\rho}$ where T is the string tension and ρ is the linear mass density of the string) are unknown. Write down an algorithm for estimating these parameters by matching the correlation function of $u(t,x)$ with the estimated correlation function based on space-time averages. Also write down the likelihood function of the parameters (c, L) in terms of Fourier series coefficients

$$v[n] = (2/L) \int_0^T \int_0^L u(t,x)sin(n\pi ct/L)sin(n\pi x/L)dtdx, T = L/c$$

From this likelihood function, devise an algorithm for estimating these two parameters using the maximum likelihood method.

[5] Let $B_1(t), B_2(t)$ be two standard independent Brownian motion processes. Consider the Brownian sheet $B(t,s) = B_1(t)B_2(s)$. Calculate its correlation function. Calculate the eigenfunctions and eigenvalues of this correlation function over the two dimensional region $[0, T_1] \times [0, T_2]$. Hence determine the KL expansion of this Brownian sheet over the state region. Demonstrate explicitly the orthogonality of the eigenfunctions.

[6] Given a zero mean stationary Gaussian random field in discrete space $X(n,m), (n,m) \in \mathbb{Z}^2$ with correlation

$$R(n-n', m-m') = \mathbb{E}(X(n,m)X(n',m'))$$

and power spectral density

$$S(\omega_1, \omega_2) = \sum_{n,m \in \mathbb{Z}} R(n,m)exp(-j(\omega_1 n + \omega_2 m))$$

define its periodogram by

$$S_{N,M}(\omega_1, \omega_2) = \frac{1}{NM} |\sum_{n=0}^{N-1} \sum_{m=0}^{M-1} X(n,m)exp(-j(n\omega_1 + m\omega_2))|^2$$

Determine the mean and covariance of $S_{N,M}(\omega_1, \omega_2)$. Show that the periodogram is anasymptotically unbiased estimator of the power spectral density but its variance does not converge to zero as $N, M \to \infty$.

[7] Given a real signal $x[n], 0 \le n \le N-1$, define its Hartley transform by

$$H(k) = N^{-1/2} \sum_{n=0}^{N-1} x[n].cos(2\pi kn/N + \phi)$$

where $\phi \in [0, 2\pi)$ is some fixed phase. Determine an algorithm for inversion, ie, recovering the sequence $x[n]$ from the sequence $x[n]$. Note that if

$$X(k) = N^{-1/2} \sum x[n] exp(-j2\pi kn/N)$$

denotes the DFT of $x[n]$, then we have

$$H(k) = Re(exp(j\phi)X(k)) = (1/2)(exp(j\phi)X(k) + exp(-j\phi)X(-k))$$

Note that using the fact that $x[n]$ is a real sequence, we get

$$X(-k) = X(k)^*$$

and hence

$$H(k) = (1/2)(exp(j\phi)X(k) + exp(-j\phi)X(k)^*)$$

Then,

$$H(-k) = (1/2)(exp(j\phi)X(-k) + exp(-j\phi)X(k))$$

and hence

$$\left(\begin{array}{c} H(k) \\ H(-k) \end{array} \right) =$$

$$(1/2) \left(\begin{array}{cc} exp(j\phi) & exp(-j\phi) \\ exp(-j\phi) & exp(j\phi) \end{array} \right) \left(\begin{array}{c} X(k) \\ X(-k) \end{array} \right)$$

and hence

$$X(k) = -j.cosec(2\phi)(exp(j\phi)H(k) - exp(-j\phi)H(-k))$$

Now using inverse DFT and this equation, we can easily invert the Hartley transform.

10.6   A remark on quantum state transmission

1.Data transfer (communication) from a single or multitransmitters to one or several receivers involves lots of losses over the channel due to channel noise. This can be combated by using appropriate coding schemes over sufficiently long source word sequences provided that the rates of data transfer are within the capacity region determined by versions of the Shannon noisy coding theorem. For example in multi-access and broadcast channels with degradation, we have the notion of a capacity region in multitimensional rate space rather than capacity and using various forms of Shannon's random coding argument, one can derive the region of the rate function for arbitrarily small error probability in recovery. Further if one uses Cq channels in which source strings are encoded in to quantum states, then the Cq (Classical-Quantum) coding theorems guarantee a much more larger degree of communication than what one would get using classical channels. The Cq capacity is defined as the maximum of

$$I(p, W) = H(\sum_x p(x)W(x)) - \sum_x p(x)H(W(x))$$

where $p(x)$ is the classical source probability distribution, $W(x)$ is the quantum state into which the source symbol x is encoded and $H(W) = -Tr(W.log(W))$ is the Von-Neumann quantum entropy of the state W. Moreover, further increase in the degrees of freedom in information transmission can be obtained by using QQ (Quantum-Quantum) channels in which a quantum state ρ_A in the Hilbert space $\mathcal{H}_A$ is directly to be communicated over a noisy quantum channel described by a TPCP map K (like a noisy Schrodinger channel). In this case, if $|x><x|$ is a purification of the state ρ_A with reference Hilbert space $\mathcal{H}_R$, then the output state is $(K \otimes I_R)(|x><x|)$ in the Hilbert space $\mathcal{H}_B \otimes \mathcal{H}_R$ and the QQ information transmitted is then

$$I(A : B) = H(\rho_A) + H(\rho_B) - H((K \otimes I_R)(|x><x|)$$

where

$$\rho_B = K(\rho_A) = Tr_R((K \otimes I_R)(|x><x|)$$

The QQ capacity of this channel is then obtained by maximizing $I(A : B)$ over all source states ρ_A.

10.7 An example of a Cq channel

Consider a quantum electromagnetic field carrying classical information from the transmitter. If θ denotes the classical information vector, then quantum electromagnetic field carrying this classical information can be expressed as

$$X^\mu(t, r) = \sum_k [f_k^\mu(t, r|\theta)a(k) + \bar{f}_k^\mu(t, r|\theta)a(k)^*]$$

where $a(k), a(k)^*, k = 1, 2, ...$ are the photons annihilation and creation operators in momentum space satisfying the canonical commutation relations

$$[a(k), a(m)^*] = \delta(k, m)$$

When this field is transmitted over a classical channel with impulse response $h(t, r|t', r')$, the received field is given by

$$Y^\mu(t, r) = \int H^\mu_\nu(t, r|t', r') X_\nu(t', r') dt' d^3 r'$$

The receiver is an atom, or an array of atoms with electrons. Assume that a complete set of observables that describe the receiver are $\xi(m), m = 1, 2, ..., p$. The Hamiltonian of the receiver in the absence of any interactions is

$$H_0(\xi(m) : m = 1, 2, ..., p)$$

The interaction Hamiltonian of the received field with the receiver system is usually described by a bilinear coupling

$$H_{int}(Y, \xi) =$$

$$\sum K(\mu, l, m, t, r) Y^\mu_l(t, r) dt d^3 r$$

and this can be expressed in the form

$$H_{int}(\theta) = \sum_{k,m} [lambda(k, m|\theta) a(k) + \lambda(k, m|\theta)^* a(k)^*] \xi(m)$$

The total Hamiltonian of the receiver and the field interacting with each other is then

$$H(\theta) = H_0(\xi) + H_F(a, a^*) + H_{int}(a, a^*, \xi|\theta)$$

The receiver initially starts in the state ρ and the field in the state σ so that the total state of the received and the field is initially $\rho \otimes \sigma$ and after time T evolves under this interaction with the received radiation field to the state $S(T|\theta)$ which can be obtained using the Dyson series for the evolution operator $U(t, s|\theta)$ with $H_0 + H_F$ as the unperturbed Hamiltonian. Note that since the receiver variables ξ and the field variables a, a^* mutually commute, it follows that H_0 and H_F also mutually commute. We can write for the joint state of the receiver and the field after time T, the expression

$$S(T|\theta) = U(T|\theta)(\rho \otimes \sigma) U(T|\theta)^*$$

and hence after time T, the receiver state is

$$\rho(T|\theta) = Tr_2[U(T|\theta)(\rho \otimes \sigma) U(T|\theta)^*]$$

The map $\theta \to \rho(T|\theta)$ defines the Cq channel, ie, an encoding of the classical information bearing sequence θ into the quantum state $\rho(T|\theta)$ to which the standard Cq theory can be applied for source string detection.

10.8 Quantum state transformation using entangled states

Speed of information transmission can be very large: Example of quantum teleportation in which we are able to transmit a d-qubit quantum state from a one place to another using a maximally entangled state of d qubits shared between the transmitter and receiver and in addition 2d classical bits of communication. Thus, entangled states form a large resource for fast quantum transmission of information.

Let A and B share a maximally entangled state

$$\sum_{i=1}^{d} |e_i \otimes e_i >$$

where $\{|e_i >\}$ is an onb for either Hilbert space. A wishes to transmit his state

$$|\psi >= \sum_{i=1}^{d} c(i)|e_i >$$

to B by exploiting his shared entangled state. He clubs this state with his shared entangled state so that the overall state of A and B becomes

$$\sum_{i} |\psi > \otimes |e_i > \otimes |e_i >= \sum_{i,j} c(j)|e_j \otimes e_i \otimes e_i >$$

With $n = 2^d$, A owns the first two components, ie 2d qubits. A applies a unitary transformation W to his 2d qubits thereby causing the overall state to become

$$\sum_{k,m,i,j} c(j)W(k,m|j,i)|e_k \otimes e_m \otimes e_i >$$

A then measures his 2d-qubit state in the onb $\{|e_i \otimes e_j >\}_{i,j=1}^{d}$ so that if he notes the outcome of this measurement as (r, s), then the overall state collapses to

$$\sum_{i,j} c(j)W(r,s|j,i)|e_r \otimes e_s > \otimes |e_i >$$

After $A's$ measurement, $B's$ state is thus given by

$$|\psi_B >= \sum_{i,j} W(r,s|j,i)c(j)|e_i >$$

A reports his measurement outcome (r, s) to B using 2d classical bits of information. B acc,ordingly applies the gate T_{rs} to his d qubits, where T_{rs} in the basis $\{|e_i >\}$ is the inverse of the matrix $((W(r,s|j,i)))_{1 \leq i,j \leq n}$, ie

$$\sum_{i} T_{rs}(m,i)W(r,s|j,i) = \delta_{m,j}$$

Then, B's state becomes

$$T_{rs}|\psi_B> = \sum_{i,j} W(r,s|j,i)c(j)T_{rs}|e_i>$$

$$= \sum_{i,j,k} W(r,s|j,i)c(j)T_{r,s}(k,i)|e_k> = \sum_{j,k} \delta_{j,k}c(j)|e_k> = \sum_j c(j)|e_j>$$

which is precisely the state that A had wished to communicate to B. Thus, d-qubits of shared entangelement plus 2d bits of classical communication can enable a user to transmit his d-qubit state to the other person.

10.9   Generation of entangled states from tensor product states

Is possible with high efficiency. A measure of the rate of generation of maximally entangled states is

$$C_e(\rho) = sup_{K_n}\{limsuplog(L_n)/n : \| \, |\Phi_n><\Phi_n| - K_n(\rho^{\otimes n}) \, \| \to 0]\}$$

This measures that maximal rate at which entangled states can be generated by quantum operations on tensor product states. The $K'_n s$ should be local quantum operations with one way classical communications.

10.10   Security in quantum communication from eavesdroppers

Let ρ_A be the state that A wishes to transmit to B. E is the eavesdropper who can tap some of the information sent by A to B. Let $|x><x|$ be a purification of ρ with reference Hilbert space $\mathcal{H}_R$. Then, the state received on $\mathcal{H}_{ABRE}$ is given by $\rho(ABRE) = (K \otimes I_R)(|x><x|)$. By partial tracing, we can calculate the final joint state $\rho(ABE)$ on ABE. The effective information transmitted from A to B after subtracting out the eavesdropped information is then given by

$$I(A:B)-I(A:E) = H(A)+H(B)-H(AB)-(H(A)+H(E)-H(AE))$$

$$= H(B)-H(E)-H(AB)+H(AE)$$

This is to be maximized over ρ_A and it will yield the capacity of the eavesdropped channel. It can be made very large as compared to the classically eavesdropped situation.

10.11   Abstract on internet of things in electromagnetics and quantum mechanics

[a] Smart devices in medicine: For example, we can design an instrument of a very small dimension that is placed on the skin of the body and this instrument will automatically generate a beep signal whenever the blood pressure of the individual crosses a certain mark, or whenever the blood sugar level of the body crosses a certain mark, or whenever the heart rate crosses a threshold. Devices may also be designed to monitor other parameters in the body like salt level in the blood haemoglobin content, or the concentration of some other chemical in the blood and send a beep signal whenever these concentrations cross certain prescribed thresholds. It should be noted that these signals will usually be very weak and hence very sensitive detectors must be designed. One way to solve this problem would be to use quantum detectors that are very sensitive to small changes in the concentration of chemicals in the blood. Quantum detectors can be designed by generating weak electromagnetic signals that would propagate through the body, get scattered according to a pattern that is dependent upon the permittivity and permeability distribution of the blood which in turn depends upon the concentration of the various chemicals in the blood. The variations in the scattered electromagnetic field will be very weak and to detect these variations, we must have atoms/molecules in the detector whose behaviour gets affected by even very weak electromagnetic fields. Such atoms and molecules will have electrons whose Hamiltonian gets altered in the presence of weak electromagnetic fields, say on the quantum scale like what one encounters in quantum electrodynamics. The quantum electromagnetic field scattered by the blood will be described by photon creation and annihilation operators modulated by appropriate functions of the space and time coordinates with the structure of these functions being dependent upon the blood composition parameters and by these fields can cause the wave functions of the electrons within the detector to get altered say by making transitions to different states. The change in the wave function is amplified by the detector and converted into electric signals which can be read using ordinary voltage or current meters. In fact, it is known that the change in the wave function will alter the electron probability density and probability current and measurements of these quantities can be made after appropriate amplification.

Chapter 11

More Problems in Classical and Quantum Information Theory

11.1 Problems

[1] If X is a Gaussian random vector with covariance R, then since the normal distribution maximizes the entropy of a probability distribution with second moment constraint, it follows that for any random variable Z having covariance R, we have

$$H(Z) \le H(X) = (n/2)log(2\pi e) + (1/2)log|R|$$

or equivalently,

$$|R| \le (2\pi e)^{-n/2}.exp(2H(Z))$$

In particular, if Z is any scalar random variable with variance σ^2, we have

$$\sigma^2 \le (2\pi e)^{-1/2}exp(2H(Z))$$

[2] Let $X = (X_1, ..., X_n)$ be a Gaussian random vector with independent components having zero mean and variances $\sigma_i^2, i = 1, 2, ..., n$. Let $V = (V_1, ..., V_n)$ be another Gaussian random vector with independent components having zero mean and variances $\alpha_i^2, i = 1, 2, ..., n$. Assume that the input variances $\{\sigma_i^2\}$ are variable subject to the total power constraint

$$\sum_{i=1}^{n} \sigma_i^2 \le nP$$

then with $Y = X + V$, we have

$$I(X:Y) = H(Y) - H(Y|X) = H(X+V) - H(V) = (1/2)\sum_i log(1 + \sigma_i^2/\alpha_i^2)$$

Maximizing this w.r.t to $\sigma_i, i = 1, 2, ..., n$ subject to the above power constraint using Lagrange multipliers based on the function

$$E(\sigma_1, ..., \sigma_n, \lambda) = \sum_i log(\sigma_i^2 + \alpha_i^2) - \lambda(\sum_i \sigma_i^2 - nP)$$

gives us

$$\sigma_i/(\sigma_i^2 + \alpha_i^2) = \lambda\sigma_i, i = 1, 2, ..., n$$

and hence for each i, either $\sigma_i = 0$ or else

$$\sigma_i^2 = \lambda^{-1} - \alpha_i^2$$

Since we require in addition that $\sigma_i^2 > 0$ and since $log(\sigma_i^2 + \alpha_i^2)$ increases with increasing σ_i^2, it follows that

$$\sigma_i^2 = max(\theta - \alpha_i^2, 0) = (\theta - \alpha_i^2)^+, i = 1, 2, ..., n, \theta = 1/\lambda$$

The Lagrange multiplier λ or equivalently θ is determined by the power constraint

$$\sum_i (\theta - \alpha_i^2)^+ = nP \; --- (1)$$

and the channel capacity of this special kind of Gaussian channel (in which the source is Gaussian with independent samples and the noise is also Gaussian with independent samples and the source and noise are independent) with power constraint is given by

$$C = max(I(X, Y)) = (1/2) \sum_i log(1 + (\theta - \alpha_i^2)^+/\alpha_i^2) \; --- (2)$$

where θ is determined by (1).

[3] Let $X(n), n \in Z$ and $V(n), n \in Z$ be mutually independent stationary Gaussian processes with power spectral densities $S_X(\omega)$ and $S_V(\omega)$ respectively. Let

$$Y(n) = X(n) + V(n)$$

Then, $Y(n), n \in Z$ is a stationary Gaussian process with power spectrum

$$S_Y(\omega) = S_X(\omega) + S_V(\omega)$$

and if $h(X), h(Y), h(V), h(Y|X)$ denote entropy rates, for example

$$h(Y) = lim_n H(Y(1), ..., Y(n))/n, h(Y|X) = H(Y(1), ..., Y(n)|X(1), ..., X(n))/n$$

etc., then the mutual information rate (per sample) between input X and output Y is given by

$$I(X, Y) = h(Y) - h(Y|X) = h(Y) - h(V) = (4\pi)^{-1} \int_{-\pi}^{\pi} log(S_Y(\omega)/S_V(\omega))d\omega$$

$$= (4\pi)^{-1} \int_{-\pi}^{\pi} log(1 + S_X(\omega)/S_V(\omega))d\omega$$

Maximizing this w.r.t. $S_X(\omega)$ subject to the power constraint

$$(2\pi)^{-1} \int_{-\pi}^{\pi} S_X(\omega)d\omega \le P$$

in the same way as in the previous problem gives us the channel capacity of the stationary Gaussian channel with power constraint:

$$C = max I(X, Y) = (4\pi) \int_{-\pi}^{\pi} log(1 + (\theta - S_V(\omega))^+/S_V(\omega))d\omega$$

where θ is chosen so that

$$\int_{-\pi}^{\pi} (\theta - S_V(\omega))^+ d\omega/2\pi = P$$

[4] Let K, K_0 be positive definite matrices. Show that the map $K \to log(|K + K_0|/|K|)$ is convex.

Solution: Let K_1, K_2 be two positive definite matrices and let $0 \le p \le 1$. Let θ be a r.v. with $P(\theta = 1) = p, P(\theta = 0) = 1 - p$. Let V_1, V_2, X, θ be independent r.v's such that $X = N(0, K_0), V_1 = N(0, K_1), V_2 = N(0, K_2)$ and define $V(\theta) = \theta V_1 + (1 - \theta)V_2$. Then $V(\theta)$ has zero mean and covariance $pK_1 + (1 - p)K_2 = K$ say. Note that $V(\theta)$ is not normal. Define $Y(\theta) = X + V(\theta), Y_1 = X + V_1, Y_2 = X + V_2$. Then $Y(\theta)$ has covariance $K_0 + K$ while Y_1, Y_2 are respectively $N(0, K_0 + K_1)$ and $N(0, K_0 + K_2)$. Now, write

$$f(K) = (1/2).log(|K_0 + K|/|K|)$$

Then,

$$I(X : Y_k) = H(Y_k) - H(V_k) = (1/2)log(|K_0 + K_k|/|K_k|), k = 1, 2$$

$$I(X; Y(\theta)) = H(Y(\theta)) - H(V(\theta))$$

On the one hand since conditioning reduces the entropy,

$$I(X : Y(\theta)|\theta) = H(X|\theta) - H(X|Y(\theta), \theta) \ge (X) - H(X|Y(\theta)) = I(X : Y(\theta))$$

while on the other,

$$I(X : Y(\theta)|\theta) = \le H(X) - H(X|Y(\theta), \theta)$$

$$= H(X) - H(X|Y_1, \theta = 1).p - H(X|Y_2, \theta = 0)(1-p)$$

$$= H(X) - H(X|Y_1).p - H(X|Y_2).(1 - p)$$

$$= p.(H(X) - H(X|Y_1)) + (1-p).(H(X) - H(X|Y_2)) = p.I(X : Y_1) + (1-p).I(X : Y_2)$$

$$= p(H(Y_1) - H(V_1)) + (1 - p).(H(Y_2) - H(V_2))$$

$$= p.f(K_1) + (1-p)f(K_2)$$

Thus,

$$I(X:Y(\theta)) \le p.f(K_1) + (1-p)f(K_2)$$

Now let U, V be two random vectors. Minimize $I(U:V)$ over all joint distributions $p(U,V)$ subject to given covariance constraint on $[U^T, V^T]$ by variational calculus. Note that $I(U:V)$ is a convex function of $p(U,V)$ and hence it will attain its minimum when its variational derivative w.r.t $p(U,V)$ equals zero. Then since

$$I(U:V) = \int p(U,V)log(p(U,V)/p_1(U)p_2(V))dUdV$$

where

$$p_1(U) = \int p(U,V)dV, p_2(V) = \int p(U,V)dU$$

we get

$$\delta I/\delta p(U,V) =$$

$$log(p(U,V))+1-log(p_1(U)p_2(V))-\int p(U,V')(1/p_1(U))dV'$$

$$-\int p(U',V)(1/p_2(V))dU'$$

$$= log(p(U,V)) + 1 - log(p_1(U)p_2(V)) - 2$$

Thus, if $I(U:V)$ is minimized w.r.t $p(U,V)$ subject to second moment constraints $\mathbb{E}((U^T,V^T)^T(U^T,V^T)) = R$. Then we would get using Lagrange multipliers,

$$p(U,V) = p_1(U)p_2(V).exp(Q(U,V))$$

where $Q(U,V)$ is a quadratic form in (U,V). If we are further given that U is Gaussian, then it would follow that

$$p(U,V) = p_2(V).exp(Q(U,V))$$

where Q is now another quadratic form in (U,V). Integrating this equation over V gives us

$$p_1(U) = \int p_2(V).exp(Q(U,V))dV$$

Since $p_1(U)$ is the exponential of a quadratic form in U it follows then that either U, V are independent in which case $I(U:V) = 0$ or else $p_2(V)$ is the exponential of a quadratic form in V, i.e., V is also Gaussian so that $p(U,V)$ becomes a joint Gaussian distribution. Note that if the second moment constraint is such that $Cov(U,V) \ne 0$, then the latter case must hold, i.e, (U,V) must be jointly Gaussian. In our case, taking $U = X, Y = Y(\theta) = X + V(\theta)$, we see that U and V are not independent since $Cov(U,V) = Cov(X)$ and hence, the latter case must hold, i.e,

$$I(X:Y(\theta)) \ge I_0$$

where I_0 is the mutual information between two jointly Gaussian random vectors (U, W) having the same joint covariance matrix as that of $(X, Y(\theta))$. This is the same as saying that (U, W) is of the form

$$U = X, W = X + V$$

where $X = N(0, K_0), V = N(0, K(\theta)), K(\theta) = Cov(V(\theta))$ and X, V are independent. For such U, W, we find that

$$I_0 = I(U : V) = H(V) - H(V|U)$$

$$= (1/2).log(|K_0 + K(\theta)|) - (1/2)log(|K(\theta)|) = f(K(\theta))$$

where
$$K(\theta) = Cov(V(\theta)) = pK_1 + (1 - p)K_2$$

[5] Channel capacity theorem, direct part and converse.

Direct part: Encode a $w \in \{1, 2, ..., 2^{nR}\}$ chosen with a uniform distribution into an input channel codeword $X^n(w)$ having pdf $p(X^n) = \Pi_i p(x_i)$ where $X^n = (x_1, ..., x_n)$. In other words, the code C is a random code so that $X^n(w), w = 1, 2, ..., w^{nR}$ are iid with distribution $p(X^n)$. Transmit each such input codeword over the channel. If w is the message, let $Y^n(w)$ be the channel output, ie, the sequence received by the receiver. Then since the noise in the channel is independent of the input, it is clear that according to this random code, $(X^n(w), Y^n(w)), w = 1, 2, ..., 2^{nR}$ are iid pairs with each pair having the distribution $p(X^n, Y^n) = \Pi_i p(x_i, y_i)$ where $p(x_i, y_i) = p(y_i|x_i)p(x_i)$. Here, $p(x)$ is some chosen input symbol distribution and $p(y|x)$ is the transition probability of the discrete memoryless channel (DMC). The decoding process is to decode Y^n as w if $(X^n(w), Y^n)$ is jointly typical for the distribution $p(X^n, Y^n) = \Pi_i p(y_i|x_i)p(x_i)$. Thus if $= 1$ is the message sent, then a decoding error occurs iff either $(X^n(1), Y^n(1))$ is not typical or else if for some $w \neq 1$, $(X^n(w), Y^n(1))$ is typical. Noting that $(X^n(1), Y^n(1))$ has the distribution $p(X^n, Y^n)$ while $X^n(w)$ and $Y^n(1)$ are independent and hence $(X^n(w), Y^n(1))$ has the distribution $p(X^n)p(Y^n)$ where $p(X^n), p(Y^n)$ are the marginals of $p(X^n, Y^n)$ and noting that the set of pairs (X^n, Y^n) that are jointly typical has a cardinality $\leq 2^{n(H(X,Y)+\epsilon)}$, it follows by the union bound that the probability of decoding error $P(n, e)$ satisfies

$$P(n, e) \leq P((X^n, Y^n) \notin A(n, \epsilon)) + .2^{nR} 2^{-n(H(X)-\epsilon)} .2^{-n(H(Y)-\epsilon)} .2^{n(H(X,Y)+\epsilon}$$

where $A(n, \epsilon)$ is the jointly typical set. Hence this probability converges to zero if

$$R < H(X) + H(Y) - H(X, Y) - 3\epsilon$$

since the probability of $(X^n, Y^n) \notin A(n, \epsilon)$ converges to zero by Chebyshev's inequality. Thus, R is achievable if there exists an input symbol probability distribution $p(x)$ such that

$$R < H(X) + H(Y) - H(X, Y)$$

Converse part: Suppose the input code rate R is achievable, ie, $P(n, e) \to 0$. The process of encoding the message to and input codeword followed by its transmission over the channel and finally followed by decoding can be described by the Markov chain

$$w \to X^n(w) \to Y^n \to \hat{w} = f(Y^n)$$

Using Fano's inequality and the data processing inequality, we have when w is uniformly distributed over $\{1, 2, ..., 2^{nR}\}$ that

$$nR = H(w) = H(w|Y^n) + I(w : Y^n) \le H(w|\hat{w}) + I(w : Y^n) \le n\epsilon(n) + I(w : Y^n)$$

$$\le n\epsilon(n) + I(X^n : Y^n) = n\epsilon(n) + H(Y^n) - H(Y^n|X^n)$$

$$\le n\epsilon(n) + \sum_{i=1}^{n} H(Y_i) - H(Y^n|X_n)$$

$$= n\epsilon(n) + \sum_{i=1}^{n} (H(Y_i) - H(Y_i|X_i))$$

$$= n\epsilon(n) + n \sum_{Ii=1}^{n} I(X_i, Y_i) \le n\epsilon(n) + n.C$$

where we only use the fact that the channel is DMC so that $H(Y^n|X^n) = \sum_{i=1}^{n} H(Y_i|X_i)$. We do not assume any specific form of the encoding process like the input X^n is iid etc.

[6] Multiple access channels, capacity region for several sources: Let the input sources be numbered as $1, 2, ..., m$. The message transmitted by the ith node is $w_i \in \{1, 2, ..., 2^{nR_i}\}, i = 1, 2, ..., m$. The message w_i is randomly encoded into the input source sequence $X_i^n(w)$ having the probability distribution $p_i(X_i^n) = \Pi_j p_i(x_{ij})$. This code is random, ie, if $w_1, ..., w_m$ are messages at the m nodes, then $(X_i^n(w_i), i = 1, 2, ..., m)$ has the distribution $\Pi_{i=1}^{m} p_i(X_i^n)$ and the sequence m-tuples $(X_i^n(w_i), i = 1, 2, ..., m)$ as $(w_1, ..., w_m)$ varies over $\times_{i=1}^{m} \{1, 2, ..., 2^{nR_i}\}$ are independent. If $(w_1, ..., w_m)$ is the message m-tuple sent, let $Y^n(w_1, ..., w_m)$ denote the received sequence. It is clear from our random code assumption that $(X_i^n(w_i), i = 1, 2, ..., m, Y^n(w_1, ..., w_m))$ as $(w_1, ..., w_m)$ varies over $\times_{i=1}^{m} \{1, 2, ..., 2^{nR_i}\}$ are mutually independent $(m+1)$ sequence tuples and that the distribution of the $(m+1)$-tuple $(X^n(w_i), i = 1, 2, ..., m, Y^n(w_1, ..., w_m))$ is $(\Pi_{i=1}^{m} p(X_i^n)) p(Y^n|X_i^n, i = 1, 2, ..., m)$. Note that the channel is assumed to be memoryless so that

$$p(Y^n|X_i^n, i = 1, 2, ..., m) = \Pi_{j=1}^{n} p(y_j|x_{ij}, i = 1, 2, ..., m)$$

where

$$X_i^n = (x_{ij}, i = 1, 2, ..., n), Y^n = (y_j, j = 1, 2, ..., n)$$

The decoding process involves decoding the message m-tuple as $(w_1, ..., w_m)$ if $(X_i^n(w_i), i = 1, 2, ..., m, Y^n)$ is jointly typical. Thus if $(w_1 = 1, ..., w_m = 1)$ is the message set sent so that the received sequence is $Y^n(1, 1, ..., 1)$, then a decoding error occurs iff either $(X_i^n(1), i = 1, 2, ..., m, Y^n(1, 1, ..., 1))$ is not typical or else if for some $(w_1, ..., w_m) \neq (1, 1, ..., 1)$, $(X_i^n(w_i), i = 1, 2, ..., m, Y^n(1, ..., 1))$ is typical.

[7] Slepian-Wolf noiseless coding/compression theorem for correlated sources.

Let $(X^n, Y^n) = \{(X_i, Y_i)\}_{i=1}^n$ be iid bivariate (generally dependent) random vectors. If we tried to compress this data using the usual noiseless Shannon coding theorem for joint sources, ie encoders of the form $(X^n, Y^n) \to h_n(X^n, Y^n)$, we would require the joint coding rate to be $R > H(X, Y)$ for asymptotic error free decoding. On the other hand, if we tried to compress individually the sources X^n and Y^n by encoders $X^n \to f_n(X^n) \in \{1, 2, ..., 2^{nR_1}\}$, and $Y^n \to g_n(Y^n) = \{1, 2, ..., 2^{nR_2}\}$ then for error free decoding, it would be sufficient to have $R_1 > H(X), R_2 > H(Y)$. The theorem of Slepian-Wolf is that, suppose we use only individual encoders $X^n \to f_n(X^n) = I_n \in \{1, 2, ..., 2^{nR_1}\}$ and $Y^n \to g_n(Y^n) = J_n \in \{1, 2, ..., 2^{nR_2}\}$, but with a joint decoder $(I_n, J_n) \to (\hat{X}^n(I_n, J_n), \hat{Y}_n(I_n, J_n))$. Then it is necessary and sufficient for asymptotic error free decoding that $R_1 > H(X|Y), R_2 > H(Y|X), R_1 + R_2 > H(X, Y)$.

Proof of the direct part, ie, achievability/existence of a code with rates satisfying the stated constraints of Slepian and Wolf. Let $P(X^n, Y^n) = \Pi_i p(x_i, y_i)$ be the distribution of (X^n, Y^n). For each X^n, Y^n generated by the joint source according to this probability distribution, choose $I_n = f_n(X^n)$ randomly in $\{1, 2, ..., 2^{nR_1}\}$ in accordance with the uniform distribution an choose $J_n = g_n(Y^n)$ randomly in $\{1, 2, ..., 2^{nR_2}\}$ again in accordance with the uniform distribution. This forms constitutes the random code. The decoding process is to choose $\hat{X}^n(I_n, J_n) = \hat{X}'^n$ and $\hat{Y}^n(I_n, J_n) = Y'^n$ if (X'^n, Y'^n) is jointly typical for the distribution $P(.,.)$ and further if $f_n(X'^n) = I_n, g_n(Y'^n) = J_n$. Assume that X^n, Y^n is the generated source word pair. Then a decoding error occurs iff either (X^n, Y^n) is not typical or else if (X^n, Y^n) is typical and there exists a pair (X'^n, Y'^n) that is typical such that $X'^n = X^n$ or $Y'^n = Y^n$ but $f_n(X'^n) = I_n = f_n(X^n)$ and $g_n(Y'^n) = J_n = g_n(Y^n)$. Thus, using the union bound, the probability of decoding error can be upper/bounded as follows:

$$P(n, e) \leq P((X^n, Y^n) \notin A(n, \epsilon)$$

$$+ \sum_{X^n, Y^n} P(X^n, Y^n) \sum_{Y' : (X^n, Y') \in A(n, \epsilon), Y' \neq Y^n} P(g_n(Y') = g_n(Y^n)|X^n, Y^n)$$

$$+ \sum_{X^n, Y^n} pP(X^n, Y^n) \sum_{X' : (X', Y^n) \in A(n, \epsilon), X' \neq X^n} P(f_n(X') = f_n(X^n))|X^n, Y^n)$$

$$\sum_{X^n, Y^n} P(X^n, Y^n) \sum_{X', Y' : (X', Y') \in A(n, \epsilon), X' \neq X^n, Y' \neq Y^n} P(f_n(X') = f_n(X^n), g_n(Y')$$

$$= g_n(Y^n)|X^n, Y^n)$$

The first terms converges to zero by the weak law of large numbers. The second term is bounded above by

$$2^{n(H(Y|X)+\epsilon)} 2^{-nR_2}$$

because if (X^n, Y') is jointly typical, then X^n is typical and then the number of Y' for which (Y', X^n) is jointly typical for such an X^n is bounded above by $2^{n(H(Y|X)+\epsilon)}$. Moreover, by our random coding scheme, for $Y' \neq Y^n$, the random variables $g_n(Y'), g_n(Y^n), X^n$ are independent and thus for a given X^n, Y^n, and $Y' \neq Y^n$, the probability that $g_n(Y') = g_n(Y^n)$ is 2^{-nR_2}. Likewise, the second term is bounded above by $2^{n(H(X|Y)+\epsilon)} \cdot 2^{-nR_1}$. Finally, the last term is bounded above by $2^{n(H(X',Y')+\epsilon)} \cdot 2^{-n(R_1+R_2)}$. (For given X^n, Y^n there are $(2^{nR_1} - 1) \cdot (2^{nR_2} - 1)$ possible choices for $f_n(X'), g_n(Y')$ for which $f_n(X') = f_n(X^n), g_n(Y') = g_n(Y^n)$ and when $X' \neq X^n, Y' \neq Y^n$, the random variables $f_n(X'), g_n(Y')$ are independent and independent of $f_n(X^n), g_n(Y^n)$ and uniformly distributed over $\{1, 2, ..., 2^{nR_1}\}$ and $\{1, 2, ..., 2^{nR_2}\}$ respectively.

Note that this probability of decoding error actually represents the average of the decoding error probability over all the ensembles of the random code. There are two sources of randomness in this averaging, the first is the randomness in the random variables X^n, Y^n output by the source and the second is in the randomness of the code f_n, g_n.

This shows that when $R_1 > H(X|Y), R_2 > H(Y|X), R_1 + R_2 > H(X, Y)$, the average decoding error probability over all random codes converges to zero and hence there exists a sequence of codes for which the decoding error probability will converge to zero. Now we must prove the converse, namely that if sequence of (deterministic) codes exists such that the decoding error probability converges to zero, then the above Slepian-Wolf bounds on the coding rates are satisfied.

Consider the coding-decoding scheme

$$X^n \to f_n(X^n) = I_n, Y^n \to g_n(Y^n) = J_n$$

The decoding scheme is

$$(I_n, J_n) \to (\hat{X}^n(I_n, J_n), \hat{Y}^n(I_n, J_n))$$

where f_n, g_n are deterministic maps from χ^n and $\mathcal{Y}^n$ into $\{1, 2, ..., 2^{nR_1}\}$ and $\{1, 2, ..., 2^{nR_2}\}$ respectively where χ is the alphabet in which X_i assumes values and $\mathcal{Y}$ is the alphabet in which Y_i assumes values. By assumption, the decoding error probability converges to zero and therefore by Fano's inequality,

$$H(X^n, Y^n | I_n, J_n) \leq H(X^n, Y^n | \hat{X}^n, \hat{Y}^n) \leq n(R_1 + R_2)P(e, n) + 1$$

with $P(e, n) \to 0$. We write $\epsilon(n)$ for any sequence that converges to zero. Thus,

$$H(X^n, Y^n | I_n, J_n) = n\epsilon(n)$$

Also since

$$H(X^n | I_n, J_n), H(Y^n | I_n, J_n) \leq H(X^n, Y^n | I_n, J_n)$$

we get

$$H(X^n | I_n, J_n), H(Y^n | I_n, J_n) \to 0$$

and so we can write

$$H(X^n|I_n, J_n) = n\epsilon(n), H(Y^n|I_n, J_n) = n\epsilon(n)$$

Then,

$$nR_1 \geq H(I_n) \geq H(I_n|Y^n) = I(I_n : X^n|Y^n) + H(I_n|X^n, Y^n)$$

$$= I(I_n : X^n|Y^n) + n\epsilon(n)$$

$$= H(X^n|Y^n) - H(X^n|I_n, Y^n) + n\epsilon(n)$$

Now,

$$H(X^n|I_n, Y^n) \leq H(X^n|I_n, J_n) = n\epsilon(n)$$

since $J_n = g_n(Y^n)$. Thus we get

$$nR_1 \geq H(X^n|Y^n) + n\epsilon(n) = nH(X|Y) + n\epsilon(n)$$

and dividing both sides by n and then letting $n \to \infty$ gives

$$R_1 \geq H(X|Y)$$

Interchanging X^n and Y^n and I_n and J_n in this argument gives us

$$R_2 \geq H(Y|X)$$

Finally,

$$R_1 + R_2 \geq H(I_n, J_n) = I(I_n, J_n : X^n, Y^n) + H(I_n, J_n|X^n, Y^n)$$

$$= I(I_n, J_n : X^n, Y^n)$$

since

$$H(I_n, J_n|X^n, Y^n) = 0$$

because $I_n = f_n(X^n), J_n = g_n(Y^n)$ for deterministic maps f_n, g_n. It follows that

$$R_1 + R_2 \geq I(I_n, J_n : X^n, Y^n) = H(X^n, Y^n) - H(X^n, Y^n|I_n, J_n)$$

$$= H(X^n, Y^n) - n\epsilon(n) = nH(X, Y) - n\epsilon(n)$$

which gives on dividing by n and letting $n \to \infty$,

$$R_1 + R_2 \geq H(X, Y)$$

This completes the proof of the converse part of the Slepian-Wolf theorem.

[8] Converse part of the achievability region for a multiple access channel with two inputs.

We have to show that if the error probability $P(n, e) \to 0$, then there exists a probability distribution Q such that (X_1, X_2, Y, Q) has a joint distribution of the form $p(q)p(x_1|q)p(x_2|q)p(y|x_1,x_2)$ with the rates R_1, R_2 satisfying the following inequalities (Note that W_1, W_2 are independent messages and hence the corresponding codewords, $X_1^n = X_1^n(W_1)$ and $X_2^n = X_2^n(W_2)$ are also independent. Note that the codes $W_k \to X_k^n(W_k), k = 1,2$ are deterministic mapping of stochastic messages Note also that each Y_i can be regarded as a function of X_{1i}, X_{2i} and an iid noise sample by nature of of the DMC. More precisely, conditioned on (X_1^n, X_2^n), the distribution of Y^n has the form $\Pi_i p(Y_i|X_{1i}, X_{2i})$). So

$$R_1 < I(X_1 : Y|X_2, Q), R_2 < I(X_2 : Y|X_1, Q), R_1 + R_2 < I(X_1, X_2 : Y|Q)$$

Now,

$$nR_1 = H(W_1) \leq H(W_1) = H(W_1|Y^n) + I(W_1 : Y^n|) = n\epsilon(n) + I(W_1 : Y^n)$$

$$\leq n\epsilon(n) + I(X_1^n : Y^n)$$

$$= n\epsilon(n) + H(X_1^n) - H(X_1^n|Y^n)$$

$$\leq n\epsilon(n) + H(X_1^n) - H(X_1^n|Y^n, X_2^n)$$

$$= n\epsilon(n) + H(X_1^n|X_2^n) - H(X_1^n|Y^n, X_2^n) = n\epsilon(n) + I(X_1^n : Y^n|X_2^n)$$

$$= n\epsilon(n) + H(Y^n|X_2^n) - H(Y^n|X_1^n, X_2^n)$$

$$\leq n\epsilon(n) + \sum_i (H(Y_i|X_2^n) - H(Y^n|X_1^n, X_2^n)($$

$$= n\epsilon(n) + \sum_i H(Y_i|X_2^n) - \sum_i H(Y_i|X_{1i}, X_{2i})$$

$$\leq n\epsilon(n) + \sum_i (H(Y_i|X_{2i}) - H(Y_i|X_{1i}, X_{2i}))$$

$$= n\epsilon(n) + \sum_i I(X_{1i} : Y_i|X_{2i})$$

Thus,

$$R_1 \leq \epsilon(n) + n^{-1} \sum_{i=1}^n I(X_{1i} : Y_i|X_{2i})$$

Define a r.v Q which assumes the values $1, 2, ..., n$ with equal probabilities of $1/n$ independently of X_1^n, X_2^n, Y^n. Then, we can write

$$n^{-1} \sum_{i=1}^n I(X_{1i} : Y_i|X_{2i}) = \sum_{i=1}^n P(Q = i)I(X_{1Q} : Y_Q|X_{2Q}, Q = i)$$

$$= I(X_{1Q} : Y_Q|X_{2Q}, Q)$$

and hence the achievability region for R_1 follows. Likewise for R_2. For $R_1 + R_2$, we have

$$n(R_1 + R_2) = H(W_1, W_2) = H(W_1, W_2|Y^n) + I(W_1, W_2 : Y^n)$$

$$= n\epsilon(n) + I(W_1, W_2 : Y^n)$$

$$\leq n\epsilon(n) + I(X_1^n, X_2^n : Y^n)$$

$$= n\epsilon(n) + H(Y^n) - H(Y^n|X_1^n, X_2^n) \leq n\epsilon(n) + \sum H(Y_i) - H(Y^n|X_1^n, X_2^n)$$

$$= n\epsilon(n) + \sum_i (H(Y_i) - H(Y_i|X_{1i}, X_{2i}))$$

$$= n\epsilon(n) + \sum_i I(X_{1i}, X_{2i} : Y_i)$$

and hence

$$R_1 + R_2 \leq \epsilon(n) + I(X_{1Q}, X_{2Q} : Y_Q|Q)$$

and this completes the converse part of the achievability proof.

Direct part of the achievability proof: The independent messages $w_1 \in \{1, 2, ..., 2^{nR_1}\} = E_1$ and $w_2 \in \{1, 2, ..., 2^{nR_2}\} = E_2$ are encoded as $X_1^n(w_1)$ and $X_2^n(w_2)$ respectively and transmitted over the channel. This code is assumed to be a random code, ie, the random vectors $(X_1^n(w_1), X_2^n(w_2)), w_1 \in E_1, w_2 \in E_2$ are assumed to be iid random vectors with the product distribution $\Pi_i p_1(x_{1i}) p_2(x_{2i}) = p_1^n(X_1^n) p_2^n(X_2^n)$. When the message (w_1, w_2) is transmitted, the receiver receives $Y^n(w_1, w_2)$ which conditioned on $(X_1^n(w_1), X_2^n(w_2))$ has the pdf $p(Y^n|X_1^n, X_2^n) = \Pi_i p(y_i|x_{1i}, x_{2i})$. Assume that $w_1 = 1, w_2 = 1$ is the message transmitted. The receiver receives $Y^n(1, 1)$ and decodes the message pair as (w_1, w_2) provided that $(X_1^n(w_1), X_2^n(w_2), Y^n(1, 1))$ is jointly typical. A decoding error occurs if wither $(X_1^n(1), X_2^n(1), Y^n(1, 1))$ is not typical, the probability of which goes to zero by the WLLN, or else if for some $(w_1, w_2) \neq (1, 1)$, $(X_1^n(w_1), X_2^n(w_2), Y^n(1, 1))$ is jointly typical. The probability of this happening is bounded above (in view of the union bound) by the sum of the probabilities of the events (a) $(X_1^n(w_1), X_2^n(1), Y^n(1, 1))$ is typical for some $w_1 \neq 1$, (b) $(X_1^n(1), X_2^n(w_2), Y^n(1, 1))$ is typical for some $w_2 \neq 1$, (c) $(X_1^n(w_1), X_2^n(w_2), Y^n(1, 1))$ is typical for some $w_1 \neq 1, w_2 \neq 1$. This sum is bounded above by the sum of

$$2^{nR_1} 2^{-nH(X_1)} . 2^{-nH(X_2, Y)} . 2^{nH(X_1, X_2, Y)} = 2^{n(R_1 - I(X_2, Y : X_1))},$$

$$2^{n(R_2 - I(X_1, Y : X_2))}$$

and

$$2^{n(R_1 + R_2)} . 2^{-nH(X_1)} . 2^{-nH(X_2)} . 2^{-nH(Y)} . 2^{nH(X_1, X_2, Y)}$$

$$= 2^{n(R_1 + R_2)} . 2^{-nI(X_1, X_2 : Y)}$$

Note that since X_1, X_2 are independent, we can write

$$I(X_2 : Y|X_1) = H(X_2|X_1) - H(X_2|Y, X_1) = H(X_2) - H(X_2|Y, X_1) = I(X_1, Y : X_2),$$

$$I(X_1 : Y|X_2) = I(X_2, Y : X_1)$$

Thus the average decoding error probability converges to zero provided

$$R_1 < I(X_1 : Y|X_2), R_2 < I(X_2 : Y|X_1), R_1 + R_2 < I(X_1, X_2 : Y)$$

for some product probability distribution $p_1(x_1)p_2(x_2)$ for (X_1, X_2) so that the joint probability distribution of (X_1, X_2, Y) becomes $p_1(x_1)p_2(x_2)p(y|x_1, x_2)$ where $p(y|x_1, x_2)$ is the transition probability distribution of the discrete memoryless multiple access channel. It is easily seen that the above achievability region for (R_1, R_2) is convex and therefore it follows that if this region is the same as the region $R_1 < I(X_1 : Y|X_2)_Q, R_2 < I(X_2 : Y|X_1, Q), R_1 + R_2 < I(X_1, X_2 : Y)_Q$ where the above mutual informations are computed using $p_1(x_1|q)p_2(x_2|q)p(y|x_1, x_2)$ for some r.v. Q and then averaged w.r.t the probability distribution of Q, ie,

$$I(X_1 : Y|X_2)_Q = I(X_2, Y : X_1|Q), I(X_2 : Y|X_1)_Q = I(X_1, Y : X_2|Q),$$

$$I(X_1, X_2 : Y)_Q = I(X_1, X_2 : Y|Q)$$

for (X_1, X_2, Y, Q) having the joint distribution $p_Q(q)p_1(x_1|q)p_2(x_2|q)p(y|x_1, x_2)$.
Remarks:
Let

$$I = (I_1, I_2, I_3), I_1 = I(X_1 : Y|X_2), I_2 = I(X_2 : Y|X_1), I_3 = I(X_1, X_2 : Y)$$

Then we write $C(I)$ for the set of all pairs (R_1, R_2) which satisfy $R_1 < I_1, R_2 < I_2, R_1 + R_2 < I_3$. $C(I)$ is the achievability region for the vector I. We see that $C(I)$ is a convex region using the fact that $I_3 \geq I_1, I_2$ because

$$I_3 - I_1 = H(Y) - H(Y|X_1, X_2) - H(Y|X_2) + H(Y|X_1, X_2) = I(Y : X_2) \geq 0$$

and likewise $I_3 - I_2 = I(Y : X_1) \geq 0$. By convexity of $C(I)$ we mean that if I' is another vector satisfying $I_3' - I_1', I_3' - I_2' \geq 0$ and $t \in [0, 1]$, then

$$C(t.I + (1 - t)I') = tC(I) + (1 - t)C(I')$$

The containment

$$tC(I) + (1 - t)C(I') \subset C(t.I + (1 - t)I')$$

is trivial. Going the other way, we observe that $C(tI + (1-t)I')$ is a convex set whose extreme points are the corresponding convex combinations of the extreme points of $C(I)$ and $C(I')$. Since a convex set is the convex hull of its extreme points, it then follows that

$$C(tI + (1 - t)I') \subset tC(I) + (1 - t)C(I')$$

proving our claim. It therefore follows that

$$C\left(\sum_q P_Q(q)I_q\right) = \sum_q P_Q(q)C(I_q)$$

where by I_q, we mean the triplet (I_{1q}, I_{2q}, I_{3q}) with $I_{1q} = I(X_1 : Y|X_2, Q = q)I_{2q} = I(X_2 : Y|X_1, Q = q), I_{3q} = I(X_1, X_2 : Y|Q = q)$. Now if $(R_1(q), R_2(q)) \in$

$C(I_q)$ for each q, then since each I_q is obtained using a product distribution $p_1(x_1|q)p_2(x_2|q)$ for (X_1, X_2), $(R_1(q), R_2(q))$ is achievable for each q. By a result that we shall prove in the next remark, it follows then that $(\sum_q P_Q(q)R_1(q), \sum_q P_Q(q)R_2(q))$ is achievable, ie, we've shown that each point in the region $\sum_q p_Q(q)C(I_q)$ is achievable. But then it follows from the above equality that $C(\sum_q P_Q(q)I_q) = C(I_Q)$ is achievable where $I_Q = \sum_q p_Q(q)I_q$.

Remark: If $R = (R_1, R_2)$ and $R' = (R'_1, R'_2)$ are achievable rate pairs and $t \in [0, 1]$, then $t.R + (1 - t)R'$ is achievable. This follows by representing the first message $w_1 \in \{1, 2, ..., 2^{nR_1}\}$ as (w_{11}, w_{12}) where $w_{11} \in \{1, 2, ..., 2^{ntR_1}\}$ and $w_{12} \in \{1, 2, ..., 2^{n(1-t)R_1}\}$ and likewise representing the second message $w_2 \in \{1, 2, ..., 2^{nR_2}\}$ as (w_{21}, w_{22}) where $w_{21} \in \{1, 2, ..., 2^{ntR_2}\}$ and $w_{22} \in \{1, 2, ..., 2^{n(1-t)R_2}\}$. Then, encode the w_{11} into the first nt bits of X_1^n and w_{21} into the first nt bits of X_2^n in an achievable way and likewise encode w_{21} into the last $n(1 - t)$ bits of X_1^n and w_{22} into the last $n(1 - t)$ bits of X_2^n again in an achievable way. This is possible since (R_1, R_2) is an achievable pair. Then if a decoding error occurs, it must occur either in the first nt bits or in the last $n(1 - t)$ bits an hence the union bound implies that the probability of decoding error occurring cannot exceed the sum of the probabilities of decoding error occurring in the first nt bits and in the last $n(1 - t)$ bits each of which converges to zero as $n \to \infty$. From this fact it is clear by induction that if $(R_1(q), R_2(q)), q \in E$ are achievable rate pairs and if p_Q is a probability distribution on Q, then $\sum_q p_Q(q)(R_1(q), R_2(q)) = (\sum_q p_Q(q)R_1(q), \sum_q p_Q(q)R_2(q)$ is achievable.

11.2 Examples of Cq data transmission

[9] Problem: Consider the Fermionic field equations for the Hartree-Fock model:

$$H(t) = (-1/2m) \int \psi_s(t, r)^* (\nabla + ieA(t, r))\psi_s(t, r)d^3r + \int V_s(r)\psi_s(t, r)^*\psi_s(t, r)d^3r$$

$$+ \int V_{ss'}(r, r')\psi_s(t, r)^*\psi_s(t, r)\psi_{s'}(t, r')^*\psi_{s'}(t, r')d^3r'$$

where summation over the repeated indices s, s' is implied. The Fermionic field satisfy the canonical anticommutation relations (CAR):

$$\{\psi_s(t, r), \psi_{s'}(t, r')^*\} = \delta(s, s')\delta^3(r - r')$$

Calculate the Heisenberg equations of motion

$$d\psi_s(t, r)/dt = i[H(t), \psi_s(t, r)]$$

using the CAR. Assume that the initial state of the Fermionic field is $\rho(0)$. After time t, what is the state of the field $\rho(t)$? How much entropy has the external vector potential field $A(t, r)$ pumped into the system assuming it to be a stochastic field. As an approximate computation, you may assume the density at time $t = 0$ to be the Gibbsian distribution $exp(-\beta H(0))/Z(\beta)$ where $H(0)$ is evaluated with $A(0, r) = 0$. After time t, the state of the system is $\mathbb{E}(exp(-\beta H(t))/Z(\beta))$. This is a special case of a more general situation in which the state of the system is of the form $\rho(t,\theta)$ where θ is a random parameter with probability distribution $F(\theta)$. The output state at time t should be taken as $\bar{\rho}(t) = \int \rho(t,\theta)dF(\theta)$. The entropy of this state is $H(\bar{\rho}(t))$. The conditional entropy of the output state given the parameter is $\int H(\rho(t,\theta))dF(\theta)$ and hence the total entropy/information transmitted from the source which generates the random variable θ at time $t = 0$ to the output at time t must be taken as

$$I(F,\rho(t,.)) = H(\bar{\rho}(t)) - \int H(\rho(t,\theta))dF(\theta)$$

$$= H(\int \rho(t,\theta)dF(\theta)) - \int H(\rho(t,\theta))dF(\theta)$$

This situation is also valid for the transmission of classical messages using a quantum electromagnetic field over a noisy channel. It can be described as follows:

Assume that the transmitter wishes to transmit a random variable θ. He encodes this message into a quantum electromagnetic field $A(t,r|\theta), \Phi(t,r|\theta)$ which he sends to his transmitter which can be regarded as a cavity resonator antenna. This cavity resonator consists of particles like electrons, positrons and photons whose state gets altered when this quantum electromagnetic field is incident upon it. After time t, the state of these particles as a result of this interaction becomes $\rho_0(t,\theta)$. This state is transmitted over a quantum noisy channel described by a TPCP map T so that the received state is

$$T(\rho_0(t,\theta)) = \rho_1(t,\theta)$$

The net information transmitted at time t over the channel is then

$$H(\int \rho_1(t,\theta)dF(\theta)) - \int H(\rho_1(t,\theta))dF(\theta)$$

and this can be maximized over all probability distributions F to yield the capacity of this Cq channel. Another way to realize this transmission of information is to consider the channel to be a Hudson-Parthasarathy noisy Schrodinger bath that generates an evolution operator $U(t)$ satisfying the qsde

$$dU(t) = (-(iH(t,\theta)+P(t,\theta))dt + \sum_i (L_i(t,\theta)dA_i(t) - M_i(t,\theta)dA_i(t)^*$$
$$+ S_i(t,\theta)d\Lambda_i(t))U(t)$$

The parameter vector θ characterizes the channel structure and requires to be estimated. Further the input system state at time $t = 0$ is $\rho_0(\phi)$ while the channel is assumed to be in a coherent state $|e(u)>$. Here, ϕ is a random parameter that characterizes the random source at the transmitter end. Both θ and ϕ require to be estimated. The state received after time t is given by

$$\rho_1(t,\theta,\phi) = Tr_2(U(t,\theta)(\rho_0(\phi) \otimes |e(u)><e(u)|)U(t,\theta)^*)$$

and this can be represented using the action of a TPCP map $T(t, \theta)$ upon the initial system state:

$$\rho_1(t, \theta, \phi) = T(t, \theta)(\rho_0(\phi))$$

From measurements on this output state, the parameters θ, ϕ require to be estimated. As usual, we can calculate the information transmitted to the receiver by the channel about the source as well as about its own parameters assuming that the source parameter ϕ has a probability distribution $F(\phi)$ while the channel parameter θ has a probability distribution $G(\theta)$ independent of the source. The net information transmitted about these parameters by the channel to the receiver is then

$$H\left(\int \rho_1(t, \theta, \phi) dF(\theta) dG(\phi)\right) - \int H(\rho_1(t, \theta, \phi)) dF(\theta) dG(\phi)$$

and we can use Cq theory based on transmitting tensor product states to estimate these parameters with good accuracy in the asymptotic limit as the number of tensor products becomes infinite.

A more general formulation of the Cq information transmission problem is via the use of quantum filtering theory developed by Belavkin based on the Hudson-Parthasarathy quantum stochastic calculus. Assume that at the source end, the classical message to be transmitted is the random vector θ. The initial state of the quantum system at this transmitter end is $\rho_{s0}(\theta)$. The state is transmitted through the channel which constitutes the bath and alters the state of the system and bath at time t to

$$\rho(t, \theta, \phi) = U(t, \phi)(\rho_{s0}(\theta) \otimes |e(u) >< e(u)|)U(t, \phi)^*$$

Here, $|e(u) >$ is the coherent state of the bath at time $t = 0$. The receiver wishes to estimate the state of the system

$$\rho_s(t, \theta, \phi) = Tr_2(\rho(t, \theta, \phi))$$

at time t from noisy non-demolition measurements. For that purpose, the transmitter also transmits and input noise process $Y_i(t)$ through the channel bath. This noise process commutes with all the system observables. At the receiver end, the receiver measures the process

$$Y_o(t) = U(t, \phi)^* Y_i(t) U(t, \phi)$$

and this forms the non-demolition process. Its time samples mutually commute and $Y_o(t)$ also commutes with the future values of the Heisenberg observables $U(s, \phi)^* X U(s, \phi), s \geq 1$ where X is any system space observable. It should be noted that the system here consists of the transmitter and the receiver while the bath is the channel. The bath interacts with the system, ie, with both the transmitter and the receiver. From measurements of $Y_o(.)$, the receiver estimates the system state $\rho_s(t, \theta, \phi)$ on a real time basis as $\hat{\rho}_s(t)$ using the Belavkin filter with some convenient initialization. Thus, the state $\hat{\rho}_s(t)$ is available to the receiver and the receiver takes measurements on this state at times $t_1 < t_2 < ... < t_N$ using a POVM to obtain measurement outcomes $a_1, ..., a_N$ with probabilities $P(a_1, ..., a_N, t_1, ..., t_N)$. These probabilities can be matched to the explicit form of the true probabilities based on the true system state $\rho_s(t, \theta, \phi) = Tr_2(U(t, \phi)(\rho_{s0}(\theta) \otimes |e(u) >< e(u)|)U(t, \phi)^*)$ Note that the receiver cannot make measurements on the true system state $\rho_s(t)$ because this state is hidden from him by noise. He can make measurements only on the filtered state $\hat{\rho}_s(t)$ as it is this state which he obtains from the noisy non-demolition measurements $Y_o(.)$. Having thus calculated the above probabilities, the receiver can obtain maximum likelihood estimates of the source and channel classical parameters ϕ, θ.

Remark: It should be noted that actually, the receiver has an explicit formula for the filtered state $\hat{\rho}_s(t, \theta)$ with him in terms of the non-demolition measurements $Y_o(s) : s \leq t$ and the initial estimate $\rho_{s0}(\theta)$ using the Belavkin filter and he can decode θ, ϕ from this formula by matching it to the explicit form of this filtered state using the formula for $U(t, \phi)$ in the expression for $Y_o(s) = U(s, \phi)^* Y_i(s)U(s, \phi), s \leq t$. This formula for θ, ϕ will thus be a function of the receiver measurements $Y_o(.)$ and the input noise process $Y_i(.)$. A final averaging over the input noise process $Y_i(.)$ can then be carried out to yield smoothened estimates of θ, ϕ.

A better way for the receiver would be to estimate θ, ϕ by using his formula for $\hat{\rho}_s(t)$ in terms of $Y_o(.)$ based on the Belavkin filter to calculate the average of some set of system observables $X_\alpha, \alpha \in I$ at time t and to match this average to the true average of these observables based on the true system density matrix at time t, namely on $\rho_s(t, \theta, \phi) = Tr_2[U(t, \phi)(\rho_{s0}(\theta) \otimes |e(u) >< e(u)|)U(t, \phi)^*]$. In this latter formalism, we can intialize the Belavkin filter with some estimate of $\rho_{s0}(\theta)$ so that it is guaranteed that as time progresses, the state estimate will eventually converge to a good estimate of the true system state.

A third method of estimating θ, ϕ would be to use the maximum likelihood method applied to the Belavkin filtered state. The Belvakin filtered state at time t $\hat{\rho}_s(t)$ is available in terms of the output measurements. If me take the output measurements at different times but do not note their outcomes, then we have available with us the Belavkin filtered state at different times prepared to take POVM measurements. We take take these measurements at different times taking into account state collapse after each measurement and allowing for Belavkin evolution between any two successive measurements and let the outcomes of these measurements at times $t_1, ..., t_N$ be $a_1, ..., a_N$ respectively. We now calculate the joint probabilities of these outcomes as follows. At time t_1, we calculate the Belavkin filtered state $\hat{\rho}_s(t_1, \theta, \phi)$ terms of θ, ϕ starting from $\rho_{s0}(\theta)$ using the explicit formula for the Belavkin filter with $Y_o(s) = Y_o(s, \phi) = U(s, \phi)^* Y_i(s)U(s, \phi), s \leq t_1$ substituted. From this formula, we can evaluate the probability of getting a_1 at time t_1. Taking into account state collapse at t_1, we calculate the Belavkin filtered state at time t_2 starting from this collapsed state at t_1 using the explicit formula for the Belavkin filter with $Y_o(s) = Y_o(s, \phi)$ substituted and then calculate the conditional probability of getting a_2 at time t_2 from measurements on this filtered state conditioned on a_1 at t_1. The process is thus iterated N times, to obtain an explicit formula for

$$P(a_1, ..., a_N, t_1, ..., t_N | \theta, \phi) = P(a_1, t_1) P(a_2, t_2 | a_1, t_1)$$
$$P(a_3, t_3 | a_1, t_1, a_2, t_2) ... P(a_N, t_N | a_1, t_1, ..., a_{N-1}, t_{N-1})$$

This formula for the joint probability is then maximized over θ, ϕ to yield maximum likelihood estimates of these parameters.

Chapter 12

Information Transmission and Compression with Distortion, Ergodic Theorem, Quantum Blackhole Physics

12.1 The individual ergodic theorem

[1] Prove the ergodic theorem along the following lines. Let $f \in L^1(\Omega, \mathcal{F}, P)$ and let $T : (\Omega, \mathcal{F}, P) \to (\Omega, \mathcal{F}, P)$ be a measure preserving transformation. We have to prove that

$$lim n^{-1} \sum_{i=0}^{n-1} f(T^i \omega)$$

exists P almost surely. First establish the maximal ergodic theorem:
[a] Let $S_n = \sum_{i=0}^{n-1} f o T^i, n = 1, 2, ..., S_0 = 0$. Define

$$M_n = max(S_k : 0 \le k \le n)$$

Then show sequentially that on the set $\{M_{n+1} > 0\}$,

$$M_{n+1} = max(0, S_1, ..., S_{n+1}) = max(S_1, ..., S_{n+1}) = f + max(0, S_1 o T, ..., S_n o T)$$

$$= f + M_n o T \le f + M_{n+1} o T$$

and hence deduce that

$$\int_\Omega M_{n+1} dP = \int_{M_{n+1} > 0} M_{n+1} dP \le \int_{M_{n+1} > 0} f dP + \int_\Omega M_{n+1} dP$$

Conclude that

$$\int_{M_n > 0} f dP \ge 0, n = 1, 2, ...$$

This is called the maximal ergodic theorem.

[b] In the previous step, write $M_n(f)$ for f to show the explicit dependence of M_n on the function f. Let E be any invariant set, ie, $T^{-1}(E) = E$. Then consider the function $g = f\chi_E$. Show that application of the maximal ergodic theorem to this function, we get

$$\int_{M_n(g)>0} g dP \geq 0$$

Show that

$$M_n(g) = M_n(f)\chi_E$$

and hence deduce that

$$\int_{\{M_n(f)>0\}\cap E} f dP \geq 0, n = 1, 2, \dots$$

[c] Now consider the set

$$E_\lambda = \{sup_{n\geq 1} n^{-1} S_n(f) > \lambda\}$$

where λ is a real number and $S_n(f) = \sum_{i=0}^{n-1} f o T^i$. Show that

$$E_\lambda = \{sup_{n\geq 1} n^{-1} S_n(f - \lambda) > 0\} = \{sup_{n\geq 1} S_n(f - \lambda) > 0\}$$

Let E be any invariant set. Show by application of the result of step [b] that

$$\int_{E_\lambda \cap E} f dP \geq \lambda . P(E \cap E_\lambda)$$

[d] Now let $-\infty < a < b < \infty$ and define the set

$$E_{a,b}(f) = \{liminf n^{-1} S_n(f) < a < b < limsup n^{-1} S_n(f)\}$$

Show that $E_{a,b}$ is an invariant set and that if $E_\lambda = E_\lambda(f), \lambda \in \mathbb{R}$ are sets as defined in [c], then

$$E_{a,b}(f) \subset E_b(f)$$

Hence, deduce using the result of [c] that

$$\int_{E_{a,b}(f)} f dP \geq b . P(E_{a,b}(f))$$

Likewise with f replaced by $-f$ and noting that

$$E_{-b,-a}(-f) = E_{a,b}(f)$$

deduce that

$$-\int_{E_{a,b}(f)} f dP = \int_{E_{-b,-a}(-f)} (-f) dP \geq -a.P(E_{a,b}(f))$$

and hence deduce that

$$(a - b)P(E_{a,b}(f)) \geq 0$$

Conclude that

$$P(E_{a,b}(f)) = 0 \forall a < b$$

and by allowing a, b to run over rationals, deduce the individual ergodic theorem of Birkhoff.

12.2   The Shannon-Mcmillan-Breiman theorem

[2] Deduce the Shannon-Mcmillan-Breiman theorem using the Martingale convergence theorem and the ergodic theorem: Let μ be a stationary ergodic measure on $A^{\mathbb{Z}}$ under the shift transformation where A is a finite alphabet. Then

$$n^{-1} log(\mu[x_n, x_{n-1}, ..., x_1] \rightarrow \bar{H}(\mu)$$

where $\bar{H}(\mu)$ is the entropy rate of the measure/stationary process defined by

$$\bar{H}(\mu) = lim H(x_n, ..., x_1)/n = H(x_0 | x_{-1}, x_{-2}, ...)$$

where

$$H(x_n, x_{n-1}, ..., x_1) = -\mathbb{E} log(\mu[x_n, ..., x_1])$$

$$= -\sum_{x_n, ..., x_1 \in A} \mu[x_n, ..., x_1] . log(\mu[x_n, ..., x_1])$$

or equivalently, using the Cesaro theorem,

$$\bar{H}(\mu) = -\mathbb{E}(log(\mu[x_0 | x_{-1}, x_{-2}, ...]))$$

Prove this theorem along the following steps:

[a] Show that

$$log(\mu[x_n, ..., x_1, x_0]) = \sum_{k=1}^{n} log(\mu[x_k | x_{k-1}, ..., x_0]) + log(\mu[x_0])$$

Define the following measurable functions on $A^{\mathbb{Z}}$

$$f_k(x) = log(\mu[x_0 | x_{-1}, ..., x_{-k}]), k = 1, 2, ...$$

Let T denote the shift transformation on $A^{\mathbb{Z}}$. Show that

$$f_k(T^k x) = log(\mu[x_k | x_{k-1}, ..., x_0]), k \geq 1$$

Show that

$$log(\mu[x_n, ..., x_1, x_0]) = \sum_{k=1}^{n} f_k(T^k x) + log(\mu[x_0])$$

[b] Define the measurable function

$$f(x) = log(\mu[x_0 | x_{-1}, x_{-2}, ...])$$

Deduce using the Martingale convergence theorem that

$$lim f_k(x) = f(x), \mu a.s.$$

[c] Use the ergodic theorem to deduce that

$$n^{-1} \sum_{k=0}^{n-1} f(T^k x) = \mathbb{E} f(x) = -\bar{H}(\mu)$$

Define the functions

$$g_N(x) = sup_{k \geq N} |f_k(x) - f(x)|, N = 1, 2, ...$$

Show that for any fixed N and all $n > N$,

$$|n^{-1} \sum_{k=0}^{n-1} f_k(T^k x) - n^{-1} \sum_{k=0}^{n-1} f(T^k x)|$$

$$\leq n^{-1} |\sum_{k=0}^{N-1} (f_k(T^k x) - f(T^k x))|$$

$$+ n^{-1} \sum_{k=N}^{n-1} g_N(T^k x)$$

Apply the ergodic theorem to g_N to deduce that

$$limsup_n |n^{-1} \sum_{k=0}^{n-1} f_k(T^k x) - n^{-1} \sum_{k=0}^{n-1} f(T^k x)|$$

$$\leq \mathbb{E}(g_N(x))$$

From step [b] and dominated convergence, deduce that

$$lim_N \mathbb{E}(g_N(x)) = \mathbb{E}(lim_N g_N(x)) = 0$$

and hence deduce the Shannon-Mcmillan-Breiman theorem for ergodic sources.

12.3   Entropy pumped by the bath into a quantum system as measured by an observer making noisy non-demolition measurements

[3] This problem is about calculating how much entropy is pumped into an atom with an electron bound to it by a random electromagnetic field.

[4] This problem is about calculating how much entropy is pumped into a quantum field consisting of electrons, positrons and photons within a cavity resonator by the surrounding bath described by creation, annihilation and conservation processes in both the Bosonic and the Fermionic domain.

The resonator has the Hamiltonian

$$H_F = \sum_k f(k)a(k)^*a(k) + g(k)b(k)^*b(k) + \sum_{klm}[(h_1(k,l,m)b(k)^*b(l)+$$

$$h_2(k,l,m)b(k)b(l) + h_3(k,l,m)b(k)^*b(l)^*)a(m) + h.c]$$

The first summation describes the field energy of the Bosons and the Fermions while the second terms describes the interaction field energy between the Fermionic current and the Bosons. The interaction Hamiltonian between the cavity field and the surrounding bath field is given by

$$H_I(t) = i\sum_a [L_a(\{a(k), a(k)^*, b(m), b(m)^*\})dA_a(t)/dt - M_a(\{a(k), a(k)^*,$$

$$b(m), b(m)^*\})^* dA_a(t)^*/dt + S_a(\{a(k), a(k)^*, b(m), b(m)^*\})d\Lambda_a(t)/dt]$$

We use the shorthand notation $\mathbf{a}, \mathbf{a}^*, \mathbf{b}, \mathbf{b}^*$ for the operator aggregates $\{a(k)\}, \{a(k)^*\}$,
$$\{b(k)\}, \{b(k)^*\}$$
respectively. Then, we can write the Hudson-Parthasarathy noisy Schrodinger equation as

$$dU(t) = [-(iH + P(\mathbf{a}, \mathbf{a}^*, \mathbf{b}, \mathbf{b}^*))dt + L_a(\mathbf{a}, \mathbf{a}^*, \mathbf{b}, \mathbf{b}^*)dA_a(t)$$

$$-M_a(\mathbf{a}, \mathbf{a}^*, \mathbf{b}, \mathbf{b}^*)dA_a(t)^* + S_a(\mathbf{a}, \mathbf{a}^*, \mathbf{b}, \mathbf{b}^*)d\Lambda_a(t)]U(t)$$

The observer wishes to measure how much entropy the bath has pumped into the cavity resonator system. For that he takes noisy non-demolition measurements on the input bath noise passed through the system cavity and uses the Belavkin filter to determine on a dynamical basis a filtered estimate of the system state. The Belavkin filtered state in the absence of Poisson noise satisfies Belavkin's stochastic Schrodinger equation when the bath is the vaccum coherent state:

$$d\rho_B(t) = \theta_0(\rho_B(t))dt + (\rho_B(t)M + M^*\rho_B(t)) - Tr(\rho_B(t)(M + M^*))\rho_B(t))(dY_o(t)$$
$$- Tr(\rho_B(t)(M + M^*))dt)$$

where θ_0 is the Hamiltonian operator with Lindblad correction terms, ie the generator of the quantum dynamical semigroup of the GKSL master equation. The entropy of the Belavkin filtered state at time t is given by

$$S_B(t) = -Tr(\rho_B(t).log(\rho_B(t))) = -Tr(\rho_B(t)Z(t))$$

where

$$Z(t) = log(\rho_B(t))$$

Application of the rule for differentiating an exponential of a matrix (a standard result in linear algebra) yields

$$\rho_B'(t) = \rho_B(t)g(ad(Z(t))(Z'(t))$$

or equivalently,

$$Z'(t) = g(ad(Z(t))^{-1}(\rho_B(t)^{-1}\rho_B'(t))$$

where

$$g(x) = (1 - exp(-x))/x = 1 - x/2! + x^2/3! + ... + (-1)^n x^n/(n+1)! + ..$$

We can using this formula, develop a Taylor expansion for $g(x)^{-1}$ in the form

$$g(x)^{-1} = \sum_{r \geq 0} c(r)x^r$$

and we get

$$Tr(\rho_B(t)Z'(t)) = \sum_{r \geq 0} c(r)Tr(\rho_B(t)ad(Z(t))^r(\rho_B(t)^{-1}\rho_B'(t)))$$

Since

$$ad(Z(t))(\rho_B(t)) = 0$$

it follows that

$$ad(Z(t))^r(\rho_B(t)^{-1}) = 0$$

and therefore by application of Leibniz rule for differentiation,

$$\rho_B(t) ad(Z(t))^r (\rho_B(t)^{-1} \rho'_B(t))) =$$

$$\rho_B(t) \rho_B(t)^{-1} ad(Z(t))^r (\rho'_B(t)) = ad(Z(t))^r (\rho'_B(t)), r = 0, 1, 2, \ldots$$

Since

$$Tr(\rho'_B(t)) = (d/dt) Tr(\rho'_B(t)) = 0$$

and for $r \geq 1$,

$$Tr(ad(Z(t))^r (\rho'_B(t))) = 0$$

because of the identity

$$Tr([A, B]) = 0$$

for any two matrices A, B, it follows that

$$Tr(\rho_B(t) Z'(t)) = 0$$

and hence the rate of entropy increase of the Belavkin filtered state is

$$S'_B(t) = -Tr(\rho'_B(t) Z(t)) = -Tr(\rho'_B(t).log(\rho_B(t)))$$

The average value of this entropy rate when the bath is in a coherent state has to be computed. For that we require to compute the expected value of $dY_o(t)$ in the coherent bath state. For example, writing

$$Y_o(t) = U(t)^* Y_i(t) U(t),$$

we find that

$$dY_o(t) = dY_i(t) + dU(t)^* dY_i(t) U(t) + U(t)^* dY_i(t) dU(t)$$

Taking

$$Y_i(t) = \sum_a (c(a) A_a(t) + \bar{c}(a) A_a(t)^* + d(a) \Lambda_a(t))$$

gives us

$$dU(t)^* dY_i(t) = U(t)^* [\bar{c}(a) L_a^* dt + d(a) S_a^* d\Lambda_a(t) + \bar{c}(a) S_a^* dA_a(t)^*]$$

and taking adjoints,

$$dY_i(t) dU(t) = [c(a) L_a dt + d(a) S_a d\Lambda_a(t) + c(a) S_a dA_a(t)] U(t)$$

so if $F(t)$ is any operator on system and bath noise space that is measurable w.r.t the Hilbert space $\mathfrak{h} \otimes \Gamma_a(\mathcal{H}_{t]})$, we get for any $|f> \in \mathfrak{h}$ where $\mathfrak{h}$ is the system Hilbert space,

$$< f \otimes e(u) | F(t) dY_o(t) | f \otimes e(u) >=$$

$$< f \otimes e(u_{t]}) | F(t) | f \otimes e(u_{t]}) > (c(a) u_a(t) + \bar{c}(a) \bar{u}_a(t) + d(a) |u_a(t)|^2) dt$$

$$+ < f \otimes e(u_{t]}) | F(t) U(t)^* (\bar{c}(a) L_a^* + d(a) S_a^* |u_a(t)|^2 + \bar{c}(a) S_a^* \bar{u}_a(t)) U(t) || f \otimes e(u_{t]}) > dt$$

$$+ < f \otimes e(u_{t]}) | F(t) U(t)^* (c(a) L_a + d(a) S_a |u_a(t)|^2 + c(a) S_a u_a(t)) U(t) | f \otimes e(u_{t]}) > dt$$

Using this formula, we can derive an expression for the average rate of entropy increase in the Belavkin filtered state.

12.4 Prove the joint convexity of the relative entropy between two probability distributions along the following steps

[a] Let $\{a_1(i)\}, \{a_2(i)\}, \{b_1(i)\}, \{b_2(i)\}$ be four probability distributions on the same finite set. let $x \in [0, 1]$ and define the probability distributions

$$a(i) = xa_1(i) + (1-x)a_2(i), b(i) = xb_1(i) + (1-x)b_2(i)$$

Show using the convexity of the function $f(x) = x \to x.log(x)$ on $\mathbb{R}_+$ after noting that

$$xb_1(i)/b(i) + (1-x)b_2(i)/b(i) = 1$$

that

$$xa_1(i).log(xa_1(i)/xb_1(i)) + (1-x)a_2(i).log((1-x)a_2(i)/(1-x)b_2(i))$$

$$= xb_1(i)f(a_1(i)/b_1(i)) + (1-x)b_2(i)f(a_2(i)/b_2(i))$$

$$= b(i)((xb_1(i)/b(i))f(a_1(i)/b_1(i)) + ((1-x)b_2(i)/b(i))f(a_2(i)/b_2(i)))$$

$$\geq b(i)(f(xb_1(i)/b(i))(a_1(i)/b_1(i)) + ((1-x)b_2(i)/b(i))(a_2(i)/b_2(i))))$$

$$= b(i)f(xa_1(i)/b(i) + (1-x)b_2(i)/b(i)) = b(i)f(a(i)/b(i))$$

Summing this inequality over i results in the desired convexity of the relative entropy function.

Remark: Let $a(i), b(i), i = 1, 2, ..., N$ be positive numbers. Then we have more generally, with

$$a = \sum_i a(i), b = \sum_i b(i)$$

the inequality,

$$\sum_i (a(i)/b).log(a(i)/b(i)) = \sum_i (b(i)/b)(a(i)/b(i)log(a(i)/b(i)))$$

$$= \sum_i (b(i)/b)f(a(i)/b(i)) \geq f(\sum_i (b(i)/b)(a(i)/b(i)))$$

$$= f(\sum_i a(i)/b) = f(a/b)$$

or equivalently,

$$\sum_i a(i).log(a(i)/b(i)) \geq a.log(a/b) = (\sum_i a(i)).log(\sum_i a(i)/\sum_i b(i))$$

This is called the log-sum inequality.

12.5 Quantum blackhole physics and the amount of information pumped by the quantum gravitating blackhole into a system of other elementary particles

The particles interacting with each other are the gravitons of the blackhole, the electrons and positrons of the Dirac field, the photons of the electromagnetic field and the scalar Klein-Gordon Bosons. Let the fields other than the graviton field be represented by the field operators $\phi_k(x), k = 1, 2, ..., N$, some of which are Bosonic and the other Fermionic. The gravitational metric field is represented by $\chi_k(x), k = 1, 2, ..., p$. The Total Hamiltonian of all these fields is

$$L_1(\chi_k, \chi_{k,\mu}) + L_2(\phi_k, \phi_{k,\mu}) + L_3(\chi_k, \phi_m, \chi_{k,\mu}, \phi_{m,\mu})$$

The first term represents the Lagrangian of the gravitational field, the second, the Lagrangian of the other fields and the third, the interaction Lagrangian between the gravitational field and the other fields. The field equations are written down for all these fields and are solved perturbatively. The zeroth order perturbation solution corresponds to solution of free wave equations for Bosons and Fermions, the solutions to which are expressible as superpositions of Bosonic and Fermionic creation and annihilation operators. The higher order perturbation terms in these field equations are accounted for by solving the field equations perturbatively. They yield the solutions to the graviton field and the other fields as higher degree polynomials in the creation and annihilation operators. The total Hamiltonian of all the fields is then expressed in terms of these creation and annihilation operators. The unperturbed Hamiltonian part in this total Hamiltonian comprises the quadratic part of the Hamiltonian of the graviton field plus the sum of free Hamiltonians of the other fields which are all quadratic forms in the respective creation and annihilation operators. The perturbing component of the total Hamiltonian is a polynomial, ie, a mixed multinomial in all creation and annihilation operators of all the fields. By perturbatively solving the Schrodinger equation for the total evolution operator of all the fields using the Dyson series expansion, we find that the zeroth order term is the free unperturbed evolution operator which is the product of the free evolution operators of the different fields. The higher order terms contain mixed terms. We can express this evolution operator as

$$U(t) = U_0(t)W(t) = U_{01}(t)U_{02}(t)W(t)$$

where $U_0(t)$ is the free unperturbed evolution operator while $W(t)$ is the perturbing term. $U_{01}(t)$ is the free evolution operator of the other fields while $U_{02}(t)$ is the free evolution operator of the gravitational field without the cubic and higher nonlinearities in its Hamiltonian.

Let the initial state of the fields be the pure state

$$|\phi> \otimes |\chi>$$

where $|\phi>$ is the initial state of the other fields and $|\chi>$ is the initial state of the graviton field. Then, after time t, the state of the other fields is

$$\rho_s(t) = Tr_2(U(t)|\phi \otimes \chi><\phi \otimes \chi|)U(t)^*)$$

In the interaction picture, the state of the other fields is given by

$$\rho_{sI}(t) = Tr_2(W(t)|\phi \otimes \chi><\phi \otimes \chi|W(t)^*)$$

Note that

$$\rho_s(t) = U_{01}(t)\rho_{sI}(t)U_{01}(t)^*$$

The entropy pumped into the other fields by the gravitational field after time t is given by

$$H(\rho_s(t)) = H(\rho_{sI}(t))$$

12.6 Direct part of the capacity theorem for relay channels

The relay system consists of a main transmitter Tx, a main receiver Rx and an intermediate relay receiver RRx followed by an intermediate relay transmitter RTx. The transmitter Tx transmits one of 2^{nR} messages $w(i)$ from the set

$$E_R = \{1, 2, ..., 2^{nR}\}$$

during the ith time block to both RRx and Rx. RRx receives the noise corrupted signal $y_1(i)$, decodes it to give an estimate of $w(i)$ and assigns it as falling within the set (or rather bin) $S(s(i+1))$. Here, $s(i) \in \{1, 2, ..., 2^{nR_0}\}$ with $R_0 < R$. $S(s), s = 1, 2, ..., 2^{nR_0}$ is a partition of the set $E_R = \{1, 2, ..., 2^{nR}\}$ into 2^{nR_0} disjoint sets. This decoded bin value, is then conveyed by RRx to RTx who transmits the decoded bin value $s(i)$ to Rx during the ith time block. Thus the main receiver Rx during the ith time block receives a noise corrupted version $y(i)$ of a combination of both the message $w(i)$ transmitted by Tx during the ith block as well as the bin value $s(i)$ of an estimate of the message $w(i-1)$ transmitted by RTx during the ith block. Rx first uses his received message $y(i)$ during the ith block to decode, ie, obtain an estimate of $s(i)$, the bin in which $w(i-1)$ (transmitted by Tx during the $(i-1)^{th}$ block along with the estimate of $s(i)$ transmitted by RTx during the ith block) fell. Then Rx uses this bin information and his received message $y(i-1)$ during the $(i-1)^{th}$ block to obtain a final estimate of $w(i-1)$. This means that at the end of the ith time block, Rx knows $w(1), ..., w(i-1)$ while RRx knows $w(1), ..., w(i)$ (more precisely, these two receivers know estimates of these messages). This is natural to expect because Rx receives signals from the relay after a delay as compared with what the relay receives.

Remark: During the ith block, Tx transmits $w(i)$ encoded as $x(w(i)|s(i))$ where $s(i)$ is the index s of that bin in which $w(i-1)$ falls, ie, $w(i-1) \in S(s(i))$ while during the ith block, RTx transmits $s(i)$ encoded as $x_1(s(i))$ (more precisely it transmits an estimate $\hat{s}(i)$ of $s(i)$ encoded as $x_1(\hat{s}(i))$). Note that $s(i)$ is determined by $w(i-1)$.

The cycle then repeats for the messages transmitted during the succeeding blocks. The main idea is that the relay receiver RRx is able to obtain a good estimate of the bin in which the previous message fell because of lower noise in his channel connecting to Tx as well as due to the fact that obtaining an accurate estimate of the bin in which the message falls is easier than obtaining an accurate estimate of the message itself. Further, the noise in the channel between RTx and Rx is smaller than the noise between Tx and Rx and hence when RTx transmits his estimate of the bin $s(i)$ in which the previous message $w(i-1)$ fell, this bin information is decoded by Rx more accurately and accurate knowledge of the bin by Rx enables him to localize the message $w(i-1)$ and hence obtain a more accurate estimate of $w(i-1)$ based on $y(i-1)$ than would have been possible based on only information received directly from Tx owing to the fact that the direct channel between Tx and Rx is higher.

In deriving the capacity of the relay channel, we adopt a random coding scheme. x denotes the coded message transmitted by Tx while x_1 that transmitted by RTx.

The encoders:

First we generate a random code for transmission by RTx. This random code assigns at random a codeword $x_1(s)$ for each $s \in E_{R_0}$ so that $x_1(s), s \in E_{R_0}$ are iid random vectors with each $x_1(s)$ distributed as $p(x_1) = \Pi_{k=1}^n p(x_{1k})$, where $x_1 = ((x_{1k}))_{k=1}^n$. Here, $p(x_k, x_{1k})$ is some chosen joint pdf for the samples of (x, x_1). Note that $(x, x_1) = \{(x_k, x_{1k}) : k = 1, 2, ..., n\}$ are iid with pdf $p(x_k, x_{1k})$. Thus, $p(x_{1k} = \sum_{x_k} p(x_k, x_{1k})$. Assign an $s \in E_{R_0}$ randomly and uniformly to each $w \in E_R$. This defines a random partition $\{S(s) : s \in E_{R_0}\}$ of E_R. For a given $s \in E_{R_0}$, choose codewords $x(w|s), w \in E_R$ so that these are all iid with pdf $p(x|x_1(s)) = \Pi_k p(x_k|x_{1k}(s))$. This completes the description of the encoder.

The decoders: [a] RRx decodes an estimate w of $w(i)$ as that message value for which $(x(w|s(i)), x_1(s(i)), y_1(i))$ is jointly typical. Here, we assume that RRx has a good estimate of the bin $s(i)$ in which $w(i-1)$ falls. The probability of an error occurring in this portion of the decoder is smaller than the sum of the probability that $(x(w(i)|s(i)), x_1(s(i)), y_1(i))$ is not typical and the probability that for some $w(i) \neq w \in E_R$, $(x(w|s(i)), x_1(s(i)), y_1(i))$ is typical. The former probability, by the weak law of large numbers, converges to zero and the latter probability is upper bounded by

$$2^{nR} . \sum_{(x,x_1,y) \in A(n,\epsilon)} P(x_1)P(x|x_1)P(y_1|x_1)$$

Here, we have used the fact that for $w \neq w(i)$, $x(w|s(i))$ is independent of $y_1(i)$ conditioned on $x_1(s(i))$ because for fixed $s(i)$, given $x_1(s(i))$ $\{x(w|s(i)), w \in E_R\}$ is an independent collection of random vectors (in particular, for $w \neq w(i)$, $x(w|s(i))$ is independent of $x(w(i)|s(i))$ conditioned on $x_1(s(i)))$ and $y_1(i)$ is a function of $Tx - RRx$ channel noise, $x(w(i)|s(i))$ and $x_1(s(i))$ and the channel has noise that is independent of the signals transmitted. Note that the Tx transmits during the ith block, the message $w(i)$ encoded as $x(w(i)|s(i))$.

Now, (x, x_1, y_1) typical implies that (x, x_1) is typical and for such (x, x_1), $P(x, x_1) = 2^{-nH(x, x_1)}$. Further, (x, x_1, y_1) typical implies that $(x_1, y_1), x_1, y_1$ are all individually typical and hence $y_1|x_1$ is conditionally typical and for such $x_1, y_1, P(y_1|x_1) = 2^{-nH(y_1|x_1)}$. Further, the number of typical triplets (x, x_1, y_1) is $2^{nH(x, x_1, y_1)}$. Thus, the above probability is bounded above by

$$2^{n(R - H(x, x_1) - H(y_1|x_1))} . 2^{nH(x, x_1, y_1)} = 2^{n(R - H(x|x_1) - H(y_1|x_1) + H(x, y_1|x_1))}$$

$$= 2^{n(R - I(x:y_1|x_1))}$$

This probability will therefore converge to zero if

$$R < I(x : y_1|x_1)$$

[b] The main receiver Rx decodes from his received signal $y(i)$ during the ith block, an estimate s of $s(i)$ provided that $(x_1(s), y(i))$ is typical. A decoding error at this place therefore occurs if either $(x_1(s(i)), y(i))$ is non typical, the probability of which converges to zero by the weak law of large numbers (note that for any w, $(x(w|s(i)), x_1(s(i)), y(i))$ has the joint pdf $p(x, x_1, y) = p(x_1)p(x|x_1)p(y|x, x_1) = p(x, x_1)p(y|x, x_1) = \Pi_k(p(x_k, x_{1k})p(y_k x_k, x_{1k})$ and hence $(x_1(s(i)), y(i))$ has the joint distribution $p(x_1, y) = p(x_1)p(y|x_1) = \sum_x p(x, x_1, y))$, or else if for some $s \neq s(i)$, $(x_1(s), y(i))$ is jointly typical. The probability of the latter event on noting that for $s \neq s(i)$, $x_1(s)$ and $y(i)$ are independent, is given by

$$\sum_{s \neq s(i), (x_1, y) \in A(n, \epsilon)} P(x_1)P(y)$$

$$\leq 2^{nR_0} . 2^{-nH(x_1)} . 2^{-nH(y)} . 2^{nH(x_1, y)} = 2^{n(R_0 - I(x_1 : y))}$$

which converges to zero provided that

$$R_0 < I(x_1 : y)$$

Note that for $s \neq s(i)$, $x_1(s)$ and $y(i)$ are independent because $y(i)$ is a function of channel noise, $x_1(s(i))$ and $x(w(i)|s(i))$. $x_1(s)$ is independent of $x_1(s(i))$ for $s \neq s(i)$ by construction of the random code and $x(w(i)|s(i))$ is independent of $\{x_1(s), s \in E_{R_0}\}$. Another way to state the same thing is that if $s \neq s(i)$, then since $x_1(s(i))$ and $x_1(s)$ are independent, given $x_1(s(i))$, the joint distribution of $(y(i), x_1(s))$ will not contain $x_1(s)$ because $y(i)$ is built out of the data $x_1(s(i))$

and the data $x(w(i)|s(i))$ transmitted by Tx namely $w(i)$ during the ith time slot while $x_1(s)$ is a random function of s which represents the data transmitted during the $(i-1)^{th}$ slot and the data transmitted during different time slots are independent by the nature of our random coding scheme. To write this explicitly, we have

$$y(i) = f(v(i), x_1(s(i)), x(w(i)|x_1(s(i))))$$

where $v(i)$ is channel noise during the ith slot. Thus,

$$P(y(i), x_1(s)|x_1(s(i))) = P(f(v(i), x_1(s(i)), x(w(i)|s(i))), x_1(s)|x_1(s(i)))$$

$$= P(f(v(i), x_1(s(i)), x(w(i)|s(i)))|x_1(s(i))).P(x_1(s))$$

for $s \neq s(i)$.

[c] The final decoding step at Rx involves Rx making use of his data $y(i-1)$ available at the $(i-1)^{th}$ block as well as his estimate of the bin $s(i)$ obtained in step [b] based on $y(i)$ (which localizes $w(i-1)$ to fall within $S(s(i))$), to obtain his estimate of the message $w(i-1)$ transmitted by Tx during the $(i-1)^{th}$ time slot. Assuming that Rx has a reliable estimate of the bin index $s(i)$, he constructs the bin $S(s(i))$ and then decodes the estimate of $w(i-1)$ as that $w \in S(s(i-1))$ for which $(x(w|s(i-1)), y(i-1))$ is conditionally typical given $x_1(s(i-1))$. A decoding error occurs if either $(x(w(i-1)|s(i-1)), y(i-1))$ conditioned on $x_1(s(i-1))$ is not typical, the probability of which goes to zero by the weak law of large numbers (because $y(i-1)$ is a function of $x_1(s(i-1)), x(w(i-1)|s(i-1))$ and channel noise and hence the joint distribution of $(y(i-1), x(w(i-1)|s(i-1)), x_1(s(i-1)))$ is $p(y|x, x_1)p(x, x_1) = p(y|x, x_1)p(x|x_1)p(x_1))$, or if for some $w(i-1) \neq w \in S(s(i)), (x(w|s(i-1)), y(i-1))$ conditioned on $x_1(s(i-1))$ is typical. This latter probability on noting that the average number of elements in $S(s(i))$ is $2^{n(R-R_0)}$ (because each element in E_R has a probability $1/2^{nR_0}$ of falling in a specified bin and the total number of elements in E_R is 2^{nR}) is upper bounded by

$$2^{n(R-R_0)} \sum_{(x,y)|\in A(n,\epsilon|x_1)} P(x|x_1)P(y|x_1)$$

$$= 2^{n(R-R_0)}2^{-nH(x|x_1)}.2^{-nH(y|x_1)}.2^{nH(x,y|x_1)}$$

$$= 2^{n(R-R_0)}2^{-nI(x:y|x_1)}$$

which will converge to zero provided that

$$R - R_0 < I(x:y|x_1)$$

12.7   An entropy inequality

[8] Problem. Show that if f is any non-random function $f : \mathbb{R}^n \to \mathbb{R}^p$ and X is any $\mathbb{R}^p$ valued random variable, then

$$H(f(X)) \leq H(X)$$

hint:

$$H(X) = H(X, f(X)) = H(f(X)) + H(X|f(X)) \geq H(f(X))$$

12.8   Entropy pumped by a random electromagnetic field and bath noise into an electron

[9] Consider Schrodinger's equation with a Lindlbad term in the position domain for mixed state dynamics:

$$i\partial_t \rho(t, r_1, r_2) = (-1/2m)(\nabla_{r_1} + ieA(t, r_1))^2$$

$$-(\nabla_{r_2} - ieA(t, r_2))^2)\rho(t, r_1, r_2) - e(\phi(t, r_1) - \phi(t, r_2))\rho(t, r_1, r_2)$$

$$+\delta. \int \theta(r_1, r_2 | r_1', r_2')\rho(t, r_1', r_2')d^3 r_1 d^3 r_2$$

Show that when $\delta = 0$, the entropy of the state remains constant in time. Evaluate using first order perturbation theory, the change in the state as a function of time t caused by the Lindlbad term.

hint: Show that when $\delta = 0$, the dynamics actually in abstract Hilbert space corresponds to

$$i\partial_t \rho(t) = [H(t), \rho(t)]$$

where

$$H(t) = (\mathbf{p} + eA(t, \mathbf{r}))^2/2m - e\phi(t, \mathbf{r})$$

where $\mathbf{r}, \mathbf{p}$ satisfy the canonical commutation relations:

$$[x_i, p_j] = i\delta_{ij}, \mathbf{r} = (x_i)_{i=1}^3, \mathbf{p} = (p_i)_{i=1}^3$$

Show that if $O(e^2)$ terms are neglected, then the above Schrodinger equation can be expressed as

$$i\partial_t \rho(t, r_1, r_2) = (-1/2m)(\nabla_{r_1}^2 - \nabla_{r_2}^2)\rho(t, r_1, r_2)$$

$$(-ie/m)((A(t, r_1), \nabla_{r_1}) + (A(t, r_2), \nabla_{r_2}))\rho(t, r_1, r_2)$$

$$-e(\phi(t, r_1) - \phi(t, r_2))\rho(t, r_1, r_2) + \delta.\theta(\rho)(t, r_1, r_2)$$

Taking e, δ as first order perturbation parameters, show that on writing

$$\rho(t, r_1, r_2) = \rho_0(r_1 - r_2) + \delta\rho(t, r_1, r_2)$$

we can write using first order perturbation theory,

$$i\partial_t \delta\rho(t, r_1, r_2) = (-1/2m)(\nabla^2_{r_1} - \nabla^2_{r_2})\delta\rho(t, r_1, r_2)$$

$$(-ie/m)((A(t, r_1), \nabla_{r_1}) + (A(t, r_2), \nabla_{r_2}))\rho_0(r_1 - r_2)$$

$$-e(\phi(t, r_1) - \phi(t, r_2))\rho_0(r_1 - r_2) + \delta.\theta(\rho_0)(r_1, r_2)$$

Note that $\rho_0(r - 1 - r_2)$ for any function $\rho_0(r)$ will satisfy the unperturbed Schrodinger equation

$$(\nabla^2_{r_1} - \nabla^2_{r_2})\rho_0(r_1 - r_2) = 0$$

Now assume that

$$A(t, r) = A_0 exp(i(\omega t - k.r)), \phi(t, r) = \phi_0.exp(i(\omega t - k.r))$$

and write

$$\delta\rho(t, r_1, r_2) = f(r_1 - r_2).exp(i(\omega t - k.(r_1 + r_2)/2))$$

Show on substitution that

$$-\omega f(r) = (-1/2m)((\nabla_r - ik/2)^2 - (\nabla_r + ik/2)^2)f(r)$$

$$-(ie/m)(exp(-ik.r/2) - exp(ik.r/2))(A_0, \nabla_r)\rho_0(r)$$

$$-e(exp(-ik.r/2) - exp(ik.r/2))\phi_0\rho_0(r) + \delta.exp(ik.(r_1 + r_2)/2)\theta(\rho_0)(r_1, r_2)$$

In order to obtain a meaningful solution to this equation for the function $f(r)$, we must assume that $\theta(\rho_0)(r_1, r_2)$ has the form

$$\theta(\rho_0)(r_1, r_2) = exp(-ik.(r_1 + r_2)/2)g(r_1 - r_2)$$

In particular, in this situation, we can ask the question, what is the difference between the entropies of the state ρ_0 which in kernel form is represented by $\rho_0(r_1 - r_2)$ and the state $\rho = \rho_0 + \delta\rho(t)$ which in kernel form is represented by $\rho_0(r_1 - r_2) + \delta\rho(t, r_1, r_2)$. The difference between these entropies will tell us how much entropy does the combination of the electromagnetic field $A(t, r), \phi(t, r)$ and the coupling to the bath defined by the Lindlbad operator $\theta(.)$ pump into the quantum system comprising the free particle of mass m.

12.9 Some problems in the detection and transmission of electromagnetic signals and image fields using quantum communication techniques

[1] This problem involves detecting a weak electromagnetic field specified by the potentials $A(t, r), \Phi(t, r)$ by exciting a system of quantum particles with this field. Let the masses of the particles in the quantum system be $m_k, k =$

$1, 2, ..., N$ and let their joint wave function be $\psi(t, r_1, ..., r_N)$. Let the charges of these particles be $-e_k, k = 1, 2, ..., N$. This wave function satisfies the Schrodinger equation

$$[\sum_{k=1}^{N}(-h^2/2m_k)(\nabla_{r_k}+ieA(t, r_k)/h)^2 - \sum_{k=1}^{N}e_k\Phi(t, r_k)]\psi(t, r_1, ..., r_N)$$

$$= ih\partial_t\psi(t, r_1, ..., r_N)$$

When classical Brownian noise is taken into account, this equation gets modified to a stochastic Schrodinger equation:

$$d_t\psi(t, r_1, ..., r_N) = [[\sum_{k=1}^{N}(ih/2m_k)(\nabla_{r_k}+ieA(t, r_k)/h)^2$$

$$+\sum_{k=1}^{N}(ie_k/h)\Phi(t, r_k)]dt - (i/h)\sum_{k=1}^{N}V_k(t, r_1, ..., r_N)dB_k(t) - - -(1)$$

$$-P(t, r_1, ..., r_k)dt]\psi(t, r_1, ..., r_N)$$

where $B_1, ..., B_N$ are independent standard Brownian motion processes and

$$P(t, r_1, ..., r_N) = (1/2)\sum_{k=1}^{N}V_k(t, r_1, ..., r_N)^2$$

is the Ito correction term that ensures that an initially normalized wave function ψ will always remain normalized, or equivalently that the evolution of the wave function is unitary. The electromagnetic fields A, Φ satisfy the Maxwell equations corresponding to a known known nonrandom charge and current source dependent upon some unknown parameters θ plus a purely random white Gaussian noise source component. We can write these Maxwell equations as

$$\partial_t^2\Phi(t, r) - \nabla^2\Phi(t, r) = \rho_0(t, r|\theta(t)) + \rho_w(t, r) - -(2)$$

$$\partial_t^2 A(t, r) - \nabla^2 A(t, r) = J_0(t, r|\theta(t)) + J_w(t, r) - -(3)$$

equations $(1), (2), (3)$ along with the parameter equations

$$d\theta(t) = dW(t)$$

where $W(.)$ is another vector valued Brownian motion process independent of $B(.)$ constitute our extended state equations for the extended state vector

$$[A, \Phi, \psi, \theta]$$

and the measurement model is obtained by considering the average value of a quantum observable X in the state $|\psi(t)>$ having position space representation

$$\psi(t, r_1, ..., r_N) = <r_1, ..., r_N|\psi(t)>$$

The measurement model is thus

$$dZ(t) = <\psi(t)|X|\psi(t)> dt + dV(t) - - - (4)$$

The problem involves estimating the extended state vector and hence the electromagnetic field on a real time basis from the measurements. More observables can also be included in these measurements. Alternately, our measurement model can be based on starting the quantum system in an initial state $|\psi(0) >$ and measuring the transition probability after time $t \mid < \psi_f|\psi(t) > |^2$ to another given state $|\psi_f >$.

[2] This problem involves compressing a time varying classical image field in the quantum domain, ie, by transforming the classical image field at each time to a pure quantum state and then fitting the parameters of Schrodinger's equation with classical noise to the measurement of the average value of an observable X made on the time evolving quantum state. Let $X_t = ((X_t(n,m)))_{1 \leq n,m \leq N}$ be the classical image field at time t. By applying a $C \rightarrow Q$ transformation, transform it into a pure quantum state $|\psi(t) > \in \mathbb{C}^{2^{N^2} \times 1}$. Then compress this pure state by encoding it as parameters in the Hamiltonian and the Lindblad operators of a stochastic Schrodinger equation. The stochastic Schrodinger equation is

$$d|\psi(t) >= (-(iH(t,\theta) + P(t,\theta))dt + \sum_k L_k(\theta)dB_k(t))|\psi(t) > - - - (1)$$

where

$$H(t,\theta) = H_0(t) + \sum_k \theta_k(t)V_k(t), L_k(\theta) = \sum_j \theta_j(t)L_{kj},$$

$$P(\theta) = (1/2) \sum_k L_k(\theta)^2$$

We take measurements on the pure quantum state $|\psi(t) >$ obtained after the $C \rightarrow Q$ transformation. This measurement process

$$dZ(t) =< \psi(t)|X|\psi(t) > dt + dV(t) - - - (2)$$

The dynamics of $\theta(t)$ is

$$d\theta(t) = dW(t) - - - (3)$$

From the extended state model $(1), (3)$ and the measurement model (2), we have to estimate $\theta(t)$ on a real time basis using the EKF. This forms the image compression algorithm based on quantum measurements taken on the state obtained by $C \rightarrow Q$ transformation of the classical image field.

[3] This problem involves exploiting the correlations between two image fields to predict the second image field based on measurements of the first image field in the quantum domain. We are given two classical image fields $X_{1t} = ((X_{1t}(n,m)))$ and $X_{2t} = ((X_{2t}(n,m)))$. We transform the first field into a quantum state $|\psi_1(t) >$ and the second field into another quantum state $|\psi_2(t) >$. We model the first state vector using a classical stochastic Schrodinger

equation with unknown parameters $\theta_1(t)$ and likewise the second state vector using a classical Schrodinger equation with another set of unknown parameters $\theta_2(t)$. These dynamics are

$$d|\psi_1(t)> = (-(iH(\theta_1) + P(\theta_1))dt + \sum_k L_k(\theta_1)dB_{1k}(t))|\psi_1(t)>,$$

$$d\theta_1(t) = dW_1(t),$$

$$d|\psi_2(t)> = (-(iH(\theta_2) + P(\theta_2))dt + \sum_k L_k(\theta_2)dB_{2k}(t))|\psi_1(t)>,$$

$$d\theta_2(t) = dW_2(t)$$

The parameters θ_1 and θ_2 are estimated on a real time basis using the EKF applied to these two state models with the measurement models being noise corrupted versions of average values of observables in the evolving states:

$$dZ_1(t) = <\psi_1(t)|X_1|\psi_1(t)> dt + dV_1(t),$$

$$dZ_2(t) = <\psi_2(t)|X_2|\psi_2(t)> dt + dV_2(t0$$

let θ_1, θ_2 denote respectively the converged parameter estimates obtained using this scheme. We repeat this process for several image pairs (X_{1t}, X_{2t}) where it is assumed that in each pair, the second field has the same relation to the first field, ie, the same sort of correlation structure exists between the parameters of the two components. This constitutes the training process. let the training set of pairs be $(X_{1t}^k, X_{2t}^k), k = 1, 2, ..., M$ and let the corresponding parameter estimate pairs be denoted by $(\theta_1^k, \theta_2^k), k = 1, 2, ..., M$ respectively. We construct a regression model between the two parameter vectors in a pair based on the hypothesis that the two image fields are correlated. The regression model is

$$\theta_1 = \Phi\theta_2 + \eta$$

Alternately, if we have just one pair of image fields, we can estimate Φ, η by minimizing

$$\int_0^T \| \theta_1(t) - \Phi\theta_2(t) - \eta \|^2 dt$$

where $\theta_1(t), \theta_2(t)$ are repectively the EKF estimates of these parameters as a function of time.

With the regression matrix and vector Φ, η obtained via a least squares fit:

$$min_{\Phi,\eta} \sum_k \| \theta_1^k - \Phi\theta_2^k - \eta \|^2$$

Our aim is then to make use of this regression model to estimate the second image field component from the first component of a fresh sample pair. This is made possible by substituting for θ_1 its expression in terms of θ_2 based on the regression model into the stochastic Schrodinger equation for $|\psi_1(t)>$ obtained from the first component image field X_{1t} after the $C \to Q$ transformation process:

$$d|\psi_1(t) >= -(iH(\Phi\theta_2(t)+\eta)+P_1(\Phi\theta_2(t)+\eta))dt$$

$$+\sum_k L_{1k}(\Phi\theta_2(t)+\eta)dB_{1k}(t))|\psi_1(t) >,$$

$$d\theta_2(t) = dW_2(t)$$

with the measurement model

$$dZ_1(t) =< \psi_1(t)|X|\psi_1(t) > dt + dV_1(t)$$

This modified extended state equations and measurement model enables us to estimate $\theta_2(t)$ based on measurements $Z_1(.)$ on the quantum state obtained from the first image field component X_{1t}.

[4] Application of Belavkin quantum filter to estimating the second component of an image field pair based on the first component. First transform the first classical image field component X_{1t} into a quantum state $|\psi_1(t) >$. Likewise transform the second component X_{2t} into the quantum state $|\psi_2(t) >$. We transmit these two states over a quantum noisy channel described by a bath in the coherent state $|\phi(u) >$. When the state $|\psi_k(t) >$ is transmitted, it gets coupled to the bath coherent state via a tensor product so that this coupled state becomes $|\psi_k(t) \otimes \phi(u) >$. Suppose that these states are sampled and transmitted at times $t_1 < t_2 < ... < t_N$. The state received at a time $t \in [t_n, t_{n+1})$ is denoted by $\rho_k(t)$. It is given by

$$\rho_k(t) = Tr_2(U(t)\rho_k(t)U(t)^*)$$

where $U(t)$ satisfies the Hudson-Parthasarathy noisy Schrodinger equation

$$dU(t) = (-(iH+P)dt+LdA(t)-L^*dA(t)^*+Sd\Lambda(t))U(t), t \in [t_n, t_{n+1}), U(t_n) = I$$

Our aim is to estimate the state $\rho_k(t)$. However, what the receiver can actually measure is only an output non-demolition process $Y_o(t)$ received at his end when an input process $Y_i(t)$ is transmitted at the transmitter end over the noisy channel bath. This output non-demolition process is given by

$$Y_o(t) = U(t)^*Y_i(t)U(t), t \in [t_n, t_{n+1})$$

From these measurements, the receiver applies the Belavkin filter to estimate $\rho_k(t)$ as $\rho_{kB}(t), t \in [t_n, t_{n+1})$. In the absence of channel noise, the true state that the receiver receives would have been

$$\rho_{k0}(t) = exp(-i(t - t_n)H)|\psi_k(t_n) >< \psi_k(t_n)|exp(i(t - t_n)H), t \in [t_n, t_{n+1})$$

By matching $\rho_{kB}(t)$ to this state $\rho_{k0}(t)$ for $t \in [t_n, t_{n+1})$, the receiver can estimate the transmitted state $|\psi_k(t_n) >$ at time t_n and hence by a $Q \to C$

transformation, he can get to know a good estimate of the classical image field X_{kt_n} at time t_n for each $k = 1, 2$. From knowledge of these two component image fields at times $t_n, n = 1, 2, ..., N$, he can design the training algorithm for calculating the regression model and hence design a filter that would estimate the second component based on the first component for a fresh image pair.

The whole point in this exercise is that the receiver can receive only a quantum state transmitted over the noisy channel bath described by the Hudson-Parthasarathy noisy Schrodinger equation and can make only non-demolition measurements and fromt this, he must be able to obtain a reliable estimate of the input transmitted quantum state and hence of the input classical image field for doing further processing.

12.10   The degraded broadcast channel

Let $w_2 \in \{1, 2, ..., 2^{nR_2}\} = E_2, w_1 \in \{1, 2, ..., 2^{nR_1}\} = E_1$. Generate iid random vectors $U^n(w_2), w_2 \in E_2$ such that each $U(w_2)$ has the pdf $p(u) = \Pi p(u_i)$ where $p(u_i) = \sum_{x_i} p(x_i, u_i)$. Conditioned on $U^n(w_2)$, generate independent $X^n(w_1, w_2), w_1 \in E_1$ with the pdf $p(X^n | U^n) = \Pi_i p(x_i | u_i)$. Here, (u, x, y_1, y_2) form a Markov chain in the order $u \to x \to y_1 \to y_2$ so that (u, x, y_1, y_2) has the pdf $p(u)p(x|u)p(y_1|x)p(y_2|y_1)$. Note that $p(y_1|x)$ and $p(y_2|y_1)$ are properties of the degraded broadcast channel. Messages w_1, w_2 are to be transmitted through the channel by encoding them into the input codeword $X^n(w_1, w_2)$. For this message pair, the receiver Rx_1 receives $Y_1^n(w_1, w_2)$ while the receiver Rx_2 receives $Y_2^n(w_1, w_2)$ so that conditioned on $X^n(w_1, w_2)$, $Y^n(w_1, w_2)$ has the distribution $p(Y^n | X^n) = \Pi_i p(y_i | x_i)$. The random code involves the generation of the iid r.v's $U^n(w_2), w_2 \in E_2$ with pdf $p(U^n) = \Pi_i p(u_i)$ and for given $w_2, U^n(w_2)$, the generation of iid r.v's $X^n(w_1, w_2), w_1 \in E_1$ having the conditional distribution $p(X^n | U^n) = \Pi_i p(x_i | u_i)$. The decoding process involves the following steps:

[1] From the received sequence Y_2^n at Rx_2, decode the second message as w_2 provided that $(U^n(w_2), Y_2^n)$ is typical ie, $\in A(\epsilon, n)$.

[2] From the received sequence Y_1^n at Rx_1, decode the first message as w_1 provided that for some $w_2 \in E_2$, $(U^n(w_2), X^n(w_1, w_2), Y_1^n)$ is typical.

Assume without loss of generality that $(w_1, w_2) = (1, 1)$ was the transmitted message pair. if a decoding error occurs, then one of the following must happen:

[a] $(U^n(1), Y_2^n(1, 1))$ is not typical.

[b] $(U^n(w_2), Y_2^n(1, 1))$ is typical for some $w_2 \neq 1$.

[c] $(U^n(1), X^n(1, 1), Y_1^n(1, 1))$ is not typical.

[d] $(U^n(w_2), X^n(w_1, w_2), Y_1^n(1, 1))$ is typical for some $(w_1, w_2) \neq (1, 1)$.

By the WLLN, the probabilities of [a] and [c] converge to zero. Since for $w_2 \neq 1$, $U^n(w_2)$ and $Y_2^n(1, 1)$ are independent, it follows that the probability of [2] is bounded above by

$$2^{nR_2} . 2^{-nH(U)} . 2^{-nH(Y_2)} . 2^{nH(U, Y_2)} = 2^{n(R_2 - I(U:Y_2))}$$

because, if $(U^n(w_2), Y_2^n(1,1))$ is typical then so are $U^n(w_2)$ and $Y_2^n(1,1)$ individually typical and hence $p(U^n(w_2)) = 2^{-nH(U)}, p(Y_2^n(1,1)) = 2^{-nH(Y_2)}$ while the number of typical $(U^n(w_2), Y_2^n(1,1))$ for fixed w_2 is $2^{nH(U,Y_2)}$ and the number of possible $w_2 \neq 1$ is $2^{nR_2} - 1$. Thus, the probability of [b] converges to zero provided that $R_2 < I(U : Y_2)$.

If [d] occurs, then either

[e] $(U^n(w_2), Y_1^n(1,1))$ is typical for some $w_2 \neq 1$,

of

[f] $(U^n(1), X^n(w_1,1), Y_1^n(1,1))$ is typical for some $w_1 \neq 1$, or

[g] $(U^n(w_2), X^n(w_1,w_2), Y_1^n(1,1))$ is typical for some $w_1 \neq 1, w_2 \neq 1$.

The probability of [e] is bounded above by

$$2^{nR_2} \cdot 2^{-nH(U)} \cdot 2^{-nH(Y_1)} \cdot 2^{nH(U,Y_1)} = 2^{n(R_2 - I(U:Y_1))}$$

which will converge to zero if $R_2 < I(U : Y_1)$ and since $U^n \to X^n \to Y_1^n \to Y_2^n$ is assumed to form a Markov chain, $I(U : Y_2) \leq I(U : Y_1)$, so $R_2 < I(U : Y_2)$ implies $R_2 < I(U : Y_1)$.

The probability of [f] is bounded above by

$$2^{nR_1} \cdot 2^{-nH(U)} \cdot 2^{-nH(X|U)} 2^{-nH(Y_1|U)} \cdot 2^{nH(U,X,Y_1)}$$

$$= 2^{n(R_1 - H(X|U) - H(Y_1|U) + H(X,Y_1|U))} = 2^{n(R_1 - I(X:Y_1|U))}$$

which will converge to zero if

$$R_1 < I(X : Y_1|U)$$

Finally, the probability of [g] is bounded above by

$$2^{n(R_1+R_2)} \cdot 2^{-nH(U,X)} \cdot 2^{-nH(Y_1)} \cdot 2^{nH(U,X,Y_1)} = 2^{-n(R_1+R_2)} \cdot 2^{-nI(U,X:Y_1)}$$

Note that

$$I(U, X : Y_1) = H(U,X) + H(Y_1) - H(U,X,Y_1)$$

$$= H(X|U) + H(Y_1) - H(X,Y_1|U) \geq H(X|U) + H(Y_1|U) - H(X,Y_1|U) = I(X : Y_1|U)$$

Also because the channel is degraded, $U \to Y_1 \to Y_2$ is a Markov chain and hence $I(U : Y_2) \leq I(U : Y_1)$ and hence

$$I(X : Y_1|U) + I(U : Y_2) \leq I(X : Y_1|U) + I(U :$$

$$Y_1) = H(Y_1|U) - H(Y_1|X,U) + H(Y_1) - H(Y_1|U) = H(Y_1) - H(Y_1|X,U) = I(U,X : Y_1)$$

Hence, the conditions $R_2 < I(U : Y_2)$ and $R_1 < I(X : Y_1|U)$ together imply $R_1 + R_2 < I(U, X : Y_1)$. Thus, we conclude that the probability of [g] will converge to zero if

$$R_2 < I(U : Y_2), R_1 < I(X : Y_1|U) - - - (1)$$

In conclusion, if there exists a r.v U such that $U \to X \to Y_1 \to Y_2$ forms a Markov chain, or equivalently, if (U, X, Y_1, Y_2) has a joint pdf of the form $p(u)p(x|u)p(y_1|c)p(y_2|y_1)$ and for this joint distribution, (1) is satisfied, then there exists an encoding scheme for the two messages being transmitted at rates R_1 and R_2 respectively such that the decoding error probability can be made arbitrarily small by choosing the length n of the codewords sufficiently large.

12.11   Rate distortion with side information

Let $(X_i, Y_i), i = 1, 2, ..., n$ be iid bivariate pairs. For each i, X_i, Y_i are mutually dependent. Let $X^n = (X_i)_{i=1}^n, Y^n = (Y_i)_{i=1}^n$. We assume the existence of a r.v. W such that the joint distribution of (W, X, Y) has the form $p(w|x)p(x|y)p(y) = p(w|x)p(x, y)$. Thus, let $W_i, i = 1, 2, ..., n$ be such that if $W^n = (W_i)_{i=1}^n$, then (W^n, X^n, Y^n) has the joint distribution

$$p(W^n, X^n, Y^n) = \Pi_i p(w_i|x_i)p(x_i|y_i)p(y_i)$$

Let
$$E_1 = \{1, 2, ..., 2^{nR_1}\}, E_2 = \{1, 2, ..., 2^{nR_2}\}$$

We define a random code $\{W^n(s) : s \in E_1\}$ of size 2^{nR_1} that encodes the 2^n sequences X^n as follows. $W^n(s), s \in E_1$ are iid random vectors with the pdf $p(W^n) = \Pi p(w_i)$ where

$$p(w_i) = \sum_{x_i} p(w_i|x_i)p(x_i) = \sum_{x_i} p(w_i, x_i)$$

Equivalently,

$$p(W^n) = \sum_{X^n} p(W^n|X^n)p(X^n) = \sum_{X^n} p(W^n, X^n)$$

The random encoding process is to select for each sequence X^n an $s \in E_1$ so that $(X^n, W^n(s))$ is typical. Here, we are assuming that there exists a unique $s \in E_1$ such that $(X^n, W^n(s))$ is typical. If there exists more than two or more such $s's$, then select the minimum value of s. If there exists no such s, then assign a given fixed $s_0 \in E_1$ to X^n. In this way, we have assigned to each sequence X^n a unique $s \in E_1$ so that our code is well defined.

The next step is to generate a random partition of E_1 into 2^{nR_2} bins in the following way. Assign each $s \in E_1$ to an $i \in E_2$ with a uniform distribution. In other words, if $s \in E_1$ falls is assigned the value $i(s) \in E_2$, then $i(s), s \in E_1$ are iid random variables, each uniformly distributed over E_2. For a given $j \in E_2$, let $B(j)$ be the subset of E_2 comprising of precisely all those $s \in E_1$ for which $i(s) = j$. Then $B(j), j = 1, 2, ..., 2^{nR_2}$ defines our random partition of E_1 into 2^{nR_2} disjoint bins. It is clear that since the probability of each $s \in E_1$ falling in a specified $B(i)$ is $1/2^{nR_2}$, the average number of elements in each $B(i)$ equals $2^{n(R_1-R_2)}$. The next stage in our random encoding process is to identify the bin $B(i)$ in which s falls and to encode X^n as i and transmit it. Thus, our random code has a size of 2^{nR_2}. We assume that $R_1 > I(W : X), R_2 = I(W : X) - I(W : Y)$. Hence, $R_1 - R_2 > I(W : Y)$.

The decoding process involves knowledge of Y^n apart from the bin number i transmitted by the encoder. The decoder selects s so that $s \in B(i)$ and $(Y^n, W^n(s))$ is typical and then estimates X^n as $\hat{X}^n = (\hat{X}_i)_{i=1}^n$ where $\hat{X}_i =$

$f(Y_i, W_i(s))$ where we assume that f and $p(w|x)$ are selected so that for a prescribed distortion D,

$$\mathbb{E}[d(x, f(y, w))] = \sum d(x, f(y, w))p(y, x, w) = \sum d(x, f(y, w))p(w|x)p(x|y)p(y) \leq D$$

and of course, for this $p(w|x)$,

$$R_1 > I(W : X), R_2 = I(W : X) - I(W : Y)$$

If a decoding error occurs, then one of the following must happen.

[a] (X^n, Y^n) is not typical.

[b] X^n is typical but there does not exist any $s' \in E_1$ for which $(X^n, W^n(s'))$ is typical.

[c] $(X^n, W^n(s))$ is typical but $(Y^n, W^n(s))$ is not typical.

[d] (X^n, Y^n) is typical but there exists another $s' \neq s$ with $s' \in B(i)$ such that $(Y^n, W^n(s'))$ is typical.

By the WLLN, the probabilities of [a] and [c] converge to zero. Note that $Y^n \to X^n \to W^n(s)$ forms a Markov chain by the way in which we have constructed the unique $s \in E_1$ for the given X^n and the way in which we have defined $W^n(s)$ as a function of X^n and the random code set. Hence, typicality of $(X^n, W^n(s))$ implies typicality of $(Y^n, X^n, W^n(s))$ implies typicality of $(Y^n, W^n(s))$.

Remark: $(X^n, W^n(s))$ is typical and (Y^n, X^n) has the distribution $p(Y^n, X^n) = \Pi_i p(y_i, x_i) = \Pi_i p(y_i|x_i)p(x_i)$. $N(a, b, c|Y^n, X^n, W^n)$ is the number of times that (a, b, c) appears in the corresponding positions of (Y^n, X^n, W^n). This is same as the number $N(a|Y^n_{b,c})$, where $Y^n_{b,c}$ is that segment of Y^n comprising those $Y'_i s$ for which (b, c) appears in the corresponding positions of (X^n, W^n), ie, for which $X_i = b, W_i = c$. Let X^n_c denote that segment of X^n for which c appears in the corresponding positions of W^n, ie, X^n_c consists of those X_i for which $W_i = c$. Then since the reversal of a Markov chain is also a Markov chain, $W^n(s) \to X^n \to Y^n$ also forms a Markov chain and hence by the law of large numbers

$$N(a, b, c|Y^n, X^n, W^n)/n = (N(a|Y^n_{b,c})/N(b|X^n_c))(N(b|X^n_c)/N(c|X^n))(N(c|W^n)/n)$$

$$\approx p_{Y|X}(a|b).p_{X|W}(b|c)p_W(c) = p_{YXW}(a, b, c)$$

proving thereby that $(Y^n, X^n, W^n(s))$ is typical. In particular, $(Y^n, W^n(s))$ is also typical. This means that the probability of [c] converges to zero.

Probability of [b] is

$$\sum_{X^n \in A(\epsilon, n)} p(X^n) \sum_{(X^n, W^n(s')) \notin A(\epsilon, n) \forall s' \in E_1} p(W^n(s'))$$

$$\sum_{X^n \in A(\epsilon, n)} p(X^n)\Pi_{s' \in E_1} \sum_{(X^n, W^n(s')) \notin A(\epsilon, n)} p(W^n(s'))$$

$$= \sum_{X^n \in A(\epsilon, n)} p(X^n)\Pi_{s' \in E_1}(1 - \sum_{(X^n, W^n(s')) \in A(\epsilon, n)} p(W^n(s')))$$

$$= \sum_{X^n \in A(\epsilon,n)} p(X^n)(1 - 2^{nH(W|X)}.2^{-nH(W)})^{2^{nR_1}}$$

$$\leq \sum_{X^n} p(X^n) exp(-2^{n(R_1 - I(W:X))}) = exp(-2^{n(R_1 - I(W:X))})$$

which will converge to zero if $R_1 > I(W : X)$.

Finally, the probability of [d] is upper bounded by

$$\sum_{Y^n} p(Y^n) \sum_{s' \in B(i), W^n(s'):(Y^n,W^n(s')) \in A(\epsilon,n)} p(W^n(s'))$$

$$\leq \sum_{Y^n} p(Y^n) 2^{-nH(W)}.2^{nH(W|Y)}.2^{n(R_1 - R_2)}$$

$$= 2^{n(R_1 - R_2 - I(W:Y))}$$

which will converge to zero if

$$R_1 - R_2 < I(W : Y)$$

ie, if

$$R_2 > R_1 - I(W : Y)$$

Since we are assuming that $R_1 > I(W : X)$, such a pair R_1, R_2 will exist iff

$$R_2 > I(W : X) - I(W : Y)$$

which proves the achievability part of the rate distortion theorem with side information.

Remark: For any $s' \in E_1$, (even if s' is dependent upon Y^n), $W^n(s')$ is independent of Y^n by the way in which we have constructed the random code $W^n(s), s \in E_1$. Further, for any given $i \in E_2$,

$$\mathbb{E}|B(i)| = 2^{n(R_1 - R_2)}$$

$$Var(|B(i)|) = 2^{nR_1}(2^{-nR_2})(1 - 2^{-nR_2})$$

These equations can also be expressed as

$$\mathbb{E}(|B(i)|/2^{n(R_1 - R_2)}] = 1, Var(|B(i)|/2^{n(R_1 - R_2)}) = 2^{-n(R_1 - R_2)}(1 - 2^{-nR_2})$$

In particular, we find that

$$|B(i)|/2^{n(R_1 - R_2)} \to 1 - - - (1)$$

in the mean square sense and it is also easily shown using the Borel-Cantelli lemma applied to the Chebyshev inequality,

$$P(||B(i)|/2^{n(R_1 - R_2)} - 1| > \epsilon) \leq Var(|B(i)|/2^{n(R_1 - R_2)})/\epsilon^2$$

$$= 2^{-n(R_1 - R_2)}(1 - 2^{-nR_2})/\epsilon^2$$

which is summable over n and hence the convergence (1) also holds in the almost sure sense. Hence

$$n^{-1}.log(|B(i)|) \to R_1 - R_2 a.s.$$

Finally, for a given typical Y^n, the number of sequences W^n for which (Y^n, W^n) is typical is $\approx 2^{n(H(W|Y)}$. This is proved by the usual typical sequence argument.

12.12 Proof of the Stein theorem in classical hypothesis testing

Under H_1, the r.v's $X(n), n = 1, 2, \ldots$ are iid with pdf $p_1(x)$ while under H_0, they are iid with pdf $p_0(x)$. The aim is to test H_1 versus H_0 based on data $X(n), n = 1, 2, \ldots, N$ with probability of false alarm $P(H_1|H_0) = \alpha(N)$ given and converging to zero while keeping for each N, the probability of miss $\beta(N) = P(H_0|H_1)$ converging to zero at the maximum possible rate. Stein's theorem is as follows: Let $Z(N)$ denote the region in which if $X^N = (X(n))_{n=1}^N$ falls, we decide H_1 and if X^N falls in $Z(N)^c$, we decide H_0. Then, for any $\delta > 0$, there exists a sequence of tests $Z(N), N \geq 1$ such that $\alpha(N) = p_0(X^N \in Z(N)) = P(H_1|H_0)$ converges to zero while

$$limsup N^{-1}.log(p_1(X^N \in Z(N)^c)) = limsup N^{-1}.log(P(H_0|H_1))$$

$$= limsup N^{-1}.log(\beta(N)) \leq -D(p_0|p_1) + \delta$$

and conversely, if for any sequence of tests $Z(N)$, $limsup \alpha(N) = \alpha < 1$, then

$$liminf N^{-1}.log(\beta(N)) \geq -D(p_0|p_1)$$

Proof: To prove the first part, we construct the optimal Neyman-Pearson test which minimizes $\beta(N)$ for a given $\alpha(N)$. This test is given by setting $Z(N)$ to be the set of all X^N for which

$$p_0(X^N)/p_1(X^N) \leq \eta(N)$$

where $\eta(N)$ is selected so that

$$p_0(Z(N)) = \alpha(N)$$

Now we can also write

$$Z(N) = \{X^N : N^{-1} \sum_{n=1}^N log(p_1(X(n)/p_0(X(n)) \geq t(N)\}$$

where

$$t(N) = log(\eta(N))/N$$

Note that

$$N^{-1} \sum_{n=1}^N log(p_1(X(n)/p_0(X(n)) = N^{-1}.log(p_1(X^N)/p_0(X^N))$$

By the law of large numbers, we have that

$$p_0(lim_N N^{-1}.log(p_1(X^N)/p_0(X^N)) = -D(p_0|p_1)) = 1$$

and

$$p_1(lim_N N^{-1}.log(p_1(X^N)/p_0(X^N)) = D(p_1|p_0)) = 1$$

Now, if we choose
$$t(N) = -D(p_0|p_1) + \delta$$
then $X^N \in Z(N)^c$ implies
$$N^{-1}log(p_1(X^N)/p_0(X^N)) \le -D(p_0|p_1) + \delta$$
implies
$$p_1(X^N) \le p_0(X^N).exp(-N(D(p_0|p_1) - \delta))$$
so that summing over $X^N \in Z(N)^c$ gives
$$p_1(Z(N)^c) \le exp(-N(D(p_0|p_1) - \delta))$$
while on the other hand,
$$p_0(Z(N)) = p_0(N^{-1}.log(p_1(X^N)/p_0(X^N)) \ge -D(p_0|p_1) + \delta) \to 0$$
as we saw by the law of large numbers. This proves the first part of Stein's lemma. For the second part, let $Z(N)$ be any sequence of tests with
$$p_0(Z(N)) \to \alpha < 1$$
Define the sets
$$T(N, \delta) = \{X^N : |N^{-1}.log(p_1(X^N)/p_0(X^N)) + D(p_0|p_1)| \le \delta\}$$
Then, $X^N \in Z(N)^c \cap T(N, \delta)$ implies
$$N^{-1}.log(p_1(X^N)/p_0(X^N)) \ge -D(p_0|p_1) + \delta$$
implies
$$p_1(X^N) \ge p_0(X^N).exp(-N(D(p_0|p_1) - \delta))$$
so that we get on summing over $X^N \in Z(N)^c \cap T(N, \delta)$,
$$p_1(Z(N)^c) \ge p_1(Z(N)^c \cap T(N, \delta))$$
$$\ge p_0(Z(N)^c \cap T(N, \delta)).exp(-N(D(p_0|p_1) - \delta)) - - (a)$$
On the other hand, by the union bound,
$$p_0(Z(N)^c \cap T(N, \delta)) \ge p_0(Z(N)^c) - p_0(T(N, \delta)^c)$$
and by the law of large numbers
$$p_0(T(N, \delta)^c) \to 0$$
while by hypothesis,
$$p_0(Z(N)^c) \to 1 - \alpha > 0$$
Thus, we get from (a),
$$liminf N^{-1}.log(p_1(Z(N)^c) \ge -D(p_0|p_1) + \delta$$
and letting $\delta \to 0$ gives us the desired conclusion of the converse part of Stein's theorem.

12.13 Source coding with side information

X^n is an iid source. Y^n is side information, $(X_i, Y_i), i = 1, 2, ...n$ is assumed to be an iid bivariate sequence. Let U be a r.v. with $p(u|y)$ prescribed. Thus,

$$p(u, x, y) = p(u|y)p(x, y) = p(u|y)p(y|x)p(x)$$

so that

$$p(U^n, X^n, Y^n) = \Pi_i p(u_i|y_i)p(y_i|x_i)p(x_i)$$

The marginal distribution of U is

$$p(u) = \sum_y p(u|y)p(y)$$

so that

$$p(U^n) = \sum_{Y^n} p(U^n|Y^n) = \Pi_i p(u_i)$$

Now let

$$E_1 = \{1, 2, ..., 2^{nR_1}\}, E_2 = \{1, 2, ..., 2^{nR_2}\}$$

The encoding process is as follows. Let $U^n(s), s \in E_2$ be iid random vectors each one having the pdf $p(U^n) = \Pi_i p(u_i)$. For each $X^n \in A^n$, let $i(X^n)$ be uniformly distributed over E_1. Thus writing

$$B(j) = \{X^n \in A^n : i(X^n) = j\}, j \in E_1$$

In this way, the set A^n of all sequences X^n has been randomly partitioned into 2^{nR_1} bins. It is clear that for any $j \in E_1$,

$$\mathbb{E}|B(j)| = a^n/2^{nR_1}, a = |A|, j \in E_1$$

More generally, for any set $E \subset A^n$,

$$\mathbb{E}(|E \cap B(j)|) = |E|/2^{nR_1}$$

The encoding process involves generating the source word X^n and choosing the bin number j to which X^n belongs, i.e, $X^n \in B(j)$. Apart from this bin information, side information from Y also has to be transmitted via an encoding process. Thus we generate Y^n (so that (X^n, Y^n) has the joint distribution $p(X^n, Y^n) = \Pi_i p(x_i, y_i)$) and choose the index $s \in E_2$ so that $(Y^n, U^n(s))$ is typical. If there are more than one such $s's$, we send the minimum s for which $(Y^n, U^n(s))$ is typical. If there is no such s, we fix some s_0 and transmit it.

Thus, the output of the encoder of the source and side information is the pair (j, s) with $i \in E_1, s \in E_2$. The decoder thus knowing (j, s), chooses that $X^n \in B(j)$ for which $(X^n, U^n(s))$ is typical.

Analysis of the error probability: If a decoding error occurs, then one of the following must happen.

[a] (X^n, Y^n) is not typical.

[b] $(Y^n, U^n(s'))$ is not typical for any $s' \in E_2$,
[c] $(X'^n, U^n(s))$ is not typical for any $X'^n \in B(j)$.
[d] $(X^n, U^n(s'))$ is typical for some $s' \neq s, s' \in B(j)$

The probability of [a] converges to zero by the weak law of large numbers.
The probability of [b] is estimated as follows.

$$P[b] = \sum_{Y^n, U^n(s'): (Y^n, U^n(s')) \notin A(\epsilon, n), \forall s' \in E_2} P((Y^n, U^n(s')))$$

$$= \sum_{Y^n \in A(\epsilon, n)} P(Y^n) \Pi_{s' \in E_2} \left(1 - \sum_{U^n(s'): (U^n(s'), Y^n) \in A(\epsilon, n)} P(U^n(s'))\right)$$

$$\leq \sum_{Y^n} P(Y^n)(1 - 2^{-nH(U)} . 2^{nH(U|Y)})^{2^{nR_2}}$$

$$= (1 - 2^{-nI(U:Y)})^{2^{nR_2}} \leq exp(-2^{n(R_2 - I(U:Y))})$$

which converges to zero provided

$$R_2 > I(U : Y)$$

Remark: Firstly, given Y^n, $U^n(s'), s' \in E_2$ are independent random vectors:

$$P(Y^n, U^n(s'), s' \in E_2) = P(Y^n)P(U^n(s'), s' \in E_2 | Y^n)$$

$$= P(Y^n)P(U^n(s'), s' \in E_2) = P(Y^n)\Pi_{s' \in E_2} P(U^n(s'))$$

Secondly, given that $(Y^n, U^n(s'))$ is typical, we have that Y^n and $U^n(s')$ are typical and that for such a Y^n, the number of $U^{n's}$ for which (Y^n, U^n) is typical is $2^{nH(U|Y)}$ while the probability of a typical $U^n(s')$ is $2^{-nH(U)}$.

Now since $(Y^n, U^n(s))$ is typical and (X^n, Y^n) has the pdf $p(X^n, Y^n) = \Pi_i p(x_i, y_i)$, it follows from the fact that $X^n \to Y^n \to U^n(s)$ is a Markov chain (because of the definite way in which $U^n(s)$ is generated from Y^n) that $(X^n, U^n(s))$ is also typical. Moreover $X^n \in B(j)$. Thus $P[c] = 0$.

Finally, if $(X^n, U^n(s'))$ is typical, so is X^n and hence $P[d]$ is upper bounded as

$$P[d] \leq |B(j) \cap A_X(n, \epsilon)| 2^{-nH(X)} . 2^{-nH(U)} . 2^{nH(X,U)}$$

$$\leq (|A_X(n, \epsilon)|/2^{nR_1}) . 2^{-nI(X:U)} = 2^{-nR_1 + nH(X) - nI(X:U)}$$

$$= 2^{-nR_1 + nH(X|U)}$$

which converges to zero provided

$$R_1 > H(X|U)$$

Remarks: Let $E_n \subset A^n$ be a sequence of non-random sets. Write $B_n(j)$ for the random set $B(j)$ showing its explicit dependence upon n. Then,

$$\mathbb{E}|E_n \cap B_n(j)| = |E_n|/2^{nR_1}$$

Further, for any $\epsilon > 0$, we have

$$P(log(|E_n \cap B_n(j)|)/n - log(|E_n|)/n + R_1 > \epsilon) =$$

$$P(|E_n \cap B_n(j)|2^{nR_1}/|E_n| > 2^{n\epsilon})$$

$$\leq (\mathbb{E}|E_n \cap B_n(j)|/|E_n|)2^{nR_1}.2^{-n\epsilon}$$

$$= 2^{-n\epsilon}$$

which is summable. Therefore, by the Borel-Cantelli lemma,

$$limsup(log(|E_n \cap B_n(j)|)/n - log(|E_n|)/n + R_1) \leq 0$$

so that

$$limsup(log(|E_n \cap B_n(j)|)/n \leq liminf log(|E_n|)/n - R_1$$

In particular, taking

$$E_n = A_X(n, \epsilon)$$

the typical sequence for X^n, we get for large n,

$$|E_n \cap B_n(j)| \leq |A_X(n, \epsilon)|2^{-nR_1} = 2^{nH(X)-nR_1}$$

Remark: By choosing the smallest $s \in E_2$ so that $(Y^n, U^n(s))$ is typical or if there does not exist such an s, choosing a fixed $s_0 \in E_2$, in effect, we are encoding the Y^n sequences using nR_2 bits. Likewise, by choosing the bin $j \in E_1$ so that $X^n \in B(j)$, we are in effect encoding the X^n sequences using nR_1 bits. Therefore, if there exists a r.v. U so that (U, Y, X) has the joint distribution $p(u|x)p(x, y)$, we can achieve any rate pair (R_1, R_2) with

$$R_1 > H(X|U), R_2 > I(Y : U)$$

for decoding X^n with negligible error probability from the codes s, j of X^n, Y^n as $n \to \infty$. This completes our discussion of source coding with side information.

12.14 Some problems on random segmentation of image fields

[1] Let $((x(n, m)))_{1 \leq n, m \leq N}$ be a random image field. Suppose we segment this image field into two disjoint subsets E, E^c. Assume that E is a random subset of I^2 where $I = \{1, 2, ..., N\}$. Such a random subset is defined by the probability distribution of $Z = |E|$, the number of elements in E and conditioned on $Z = r$, the probability that E will comprise of precisely the elements $(n_k, m_k), k = 1, 2, ..., r$. Let us denote this latter probability by

$$P_r((n_k, m_k) : k = 1, 2, ..., r|$$

and

$$Q(r) = P(Z = r)$$

Obviously we have

$$\sum_{1 \le n_k, m_k \le N, k=1,2,\ldots,r} P_r((n_k, m_k) : k = 1, 2, \ldots, r) = 1$$

and

$$\sum_{r=0}^{N_2} Q(r) = 1$$

The functions

$$P_r : (I^2)^r \to [0,1], r = 1, 2, \ldots, N^2$$

and

$$Q : \{1, 2, \ldots, N^2\} \to [0,1]$$

completely specify the statistics of the random set E. Suppose we wish to compute the statistics of the mean intensity of the set of pixels in E. This mean intensity is given by

$$f(E) = Z^{-1} \sum_{k=1}^{Z} X(n_k, m_k)$$

its expected value is given by

$$\mathbb{E}(f(E)) = \sum_{r=1}^{N^2} Q(r) r^{-1} \sum_{1 \le l \le r} \sum_{1 \le n_k, m_k \le N, k=1,2,\ldots,r} P_r((n_k, m_k) : k = 1, 2, \ldots, r) X(n_l, m_l)$$

Note that P_r is symmetric under all permutations, ie, for each $r = 1, 2, \ldots, N^2$ and each permutation σ of $\{1, 2, \ldots, r\}$, we have

$$P_r((n_{\sigma k}, m_{\sigma k}), 1 \le k \le r) = P_r((n_k, m_k), 1 \le k \le r)$$

This is because P_r is a function of subset of $(I^2)^r$ and a subset does not depend upon the order in which its elements are written down. Thus we simplify the expression for the expected mean intensity to be

$$\mathbb{E}(f(E)) = \sum_{r=1}^{N^2} Q(r) r^{-1} \sum_{1 \le n_k, m_k \le N, k=1,2,\ldots,r} X(n_1, m_1) P_r((n_k, m_k), 1 \le k \le r)$$

$$= \sum_{n,m=1}^{N^2} X(n, m) Q_1((n, m))$$

where

$$Q_1(n_1, m_1) = \sum_{r=1}^{N^2} Q(r) r^{-1} \sum_{1 \le n_k, m_k \le N, k=2,3,\dots,r} P_r((n_1, m_1), \dots, (n_r, m_r))$$

Note that the quantity

$$\tilde{P}_r(n, m) \sum_{n_2, m_2, \dots, n_r, m_r} P_r((n_1, m_1), \dots, (n_r, m_r))$$

equals the probability that one of the pixels in the set E equals (n_1, m_1) conditioned on the event that $|E| = r$. In terms of this function, we can write

$$Q_1(n, m) = \sum_{r=1}^{N^2} Q(r) r^{-1} \tilde{P}_r(n, m)$$

An elementary example of a random segmentation which appears in proving many kinds of coding theorems is the following: We assign each pixel (n, m) in the image a value 1 or 0 with probability p and $1 - p$ respectively independently of the others. This means that we define N^2 iid Bernoulli random variables $i(n, m), n, m = 1, 2, \dots, N$ such that

$$P(i(n, m) = 1) = p, P(i(n, m) = 0) = q = 1 - p$$

Then the random set E is defined to be the set of all pixels (n, m) that have been assigned the value 1, ie,

$$E = \{(n, m) : i(n, m) = 1\}$$

It is easy to see that

$$|E| = \sum_{n, m=1}^{N} i(n, m)$$

and hence $|E|$ is a binomial random variable with parameters (N^2, p). Thus,

$$P(|E| = r) = \binom{N^2}{r} p^r q^{N^2 - r}, 0 \le r \le N^2$$

and

$$\mathbb{E}(|E|^m) = \sum_{r=0}^{N^2} r^m P(|E| = r)$$

The moment generating function of $|E|$ is

$$\phi(z) = \mathbb{E}(z^{|E|}) = (pz + q)^{N^2}$$

Suppose that the image intensity values in this example $X(n, m), n, m = 1, 2, ..., N$ are iid random variables each with a probability density $g(x)$. Then the mean intensity of the pixels in the set E is given by

$$f(E) = |E|^{-1} \sum_{(n,m) \in E} X(n, m) = |E|^{-1} \sum_{n,m=1}^{N^2} i(n, m)X(n, m)$$

$$= \frac{\sum_{n,m} i(n, m)X(n, m)}{\sum_{n,m} i(n, m)}$$

Note that the set of r.v's $\{X(n, m) : 1 \leq n, m \leq N\}$ is independent of the set of random variables $\{i(n, m) : 1 \leq n, m \leq N\}$. Conditioned on the event that $|E| = r$, it is easy to see that $f(E)$ has the moment generating function $(\int exp(sx/r)g(x)dx)^r$ and hence it follows immediately that the moment generating function of the mean intensity $f(E)$ is given by

$$M(s) = \mathbb{E}(exp(sf(E))) = \sum_{\binom{N^2}{r}} p^r q^{N^2-r} \binom{N^2}{r} (\int exp(sx/r)g(x)dx)^r$$

The moment generating function of the total intensity $|E|f(E)$ of the image pixels within E has a simpler expression:

$$M_T(s) = \mathbb{E}(exp(s|E|f(E))) = (p \int exp(sx)g(x)dx + q)^{N^2}$$

From these expressions, all the moments of $f(E)$ and $|E|f(E)$ can be evaluated. A less trivial example is the situation in which the $i(n, m)'s$ are not independent but we specify their joint moments

$$M_r((n_k, m_k), k = 1, 2, ..., r) = \mathbb{E}(\Pi_{k=1}^r i(n_k, m_k))$$

where $(n_k, m_k), k = 1, 2, ..., r$ are distinct ordered pairs in I^2. We assume that these moments are symmetric under all permutations of the ordered pairs. Then, conditioned on $|E| = r$, we have that the moment generating function of $f(E)$ is again given by

$$(\int exp(sx/r)g(x)dx)^r$$

but the probability $P(|E| = r)$ cannot be given a simple expression like the binomial distribution. A direct way would be to compute

$$\mathbb{E}(|E|^r f(E)^r) = \mathbb{E}((\sum_{n,m} i(n, m)X(n, m))^r)$$

$$= \sum_{n_1,m_1,...,n_r,m_r} \mathbb{E}(\Pi_{k=1}^r i(n_k, m_k)))\mathbb{E}(\Pi_{k=1}^r X(n_k, m_k))$$

$$= \sum_{n_1,m_1,...,n_r,m_r} M_r((n_k, m_k), k = 1, 2, ..., r)\mathbb{E}(\Pi_{k=1}^r X(n_k, m_k))$$

This expression for the moments of the total intensity $|E|f(E)$ of the pixels within E is very general, it does not even require the r.v.s $X(n,m), n, m = 1, 2, ..., N$ to be independent. The only assumption for this expression to be valid is that the set of random variables $\{i(n,m) : n, m = 1, 2, ..., N\}$ should be independent of the set of random variables $\{X(n,m) : n, m = 1, 2, ..., N\}$. In other words, the random set E should be independent of the set $\{X(n,m) : n, m = 1, 2, ..., N\}$.

Now consider the more general problem of generating a random partition into r disjoint subsets of an image field $X(n,m), 1 \leq n, m \leq N$. This involves assigning a random variable $i(n,m)$ to the $(n,m)^{th}$ pixel where $i(n,m) \in \{1, 2, ..., r\}$. Then, the random sets $B(j), j = 1, 2, ..., r$ defined by

$$B(j) = \{(n,m) : i(n,m) = j\}, j = 1, 2, ..., r$$

form a random partition of $I^2 = \{(n,m) : 1 \leq n, m \leq N\}$.

12.15 The Shannon code

Let the source alphabet be

$$A = \{1, 2, ..., N\}$$

Let $p(x), x \in A$ be probability distribution on the source alphabet. Define

$$F(x) = \sum_{y \leq x} p(y), x \in A$$

ie, F is the cumulative probability distribution function of the source r.v. Define

$$\bar{F}(x) = \sum_{y < x} p(y) + p(x)/2 == F(x-1) + p(x)/2 = F(x) - p(x)/2, x \in A$$

Clearly, since we are assuming that $p(x) > 0 \forall x \in A$, it follows that

$$\bar{F}(x) \in (F(x-1), F(x))$$

Let $[u]$ denote the least integer $\geq u$ for any real number u. Thus, if

$$n < u \leq n+1$$

for an integer n, then

$$[u] = n+1$$

Define

$$l(x) = [log(1/p(x))] + 1, x \in A$$

Clearly,

$$l(x) \geq -log(p(x)) + 1$$

and hence

$$2^{-l(x)} \le p(x)/2 = \bar{F}(x) - F(x-1) = F(x) - \bar{F}(x)$$

Now,

$$\bar{F}(x) = F(x) - p(x)/2 \le F(x) - 2^{-l(x)},$$

Let $[u]_l$ denote the truncation of the binary expansion of the positive real number u to l binary places. Clearly, if the binary expansion of u is

$$u = 0.u_1 u_2 u_3 ...$$

then

$$u = 0.u_1 u_2 ... u_l ... \le 0.u_1 ... u_l 111... = 0.u_1 ... u_l + 1/2^l = [u]_l + 1/2^l$$

Thus,

$$u - 1/2^l \le [u]_l$$

and hence,

$$F(x) - 1/2^{l(x)} \le [\bar{F}(x)]_{l(x)} \le \bar{F}(x) = F(x-1) + p(x)/2 = F(x) - p(x)/2 < F(x)$$

In other words,

$$[\bar{F}(x)]_{l(x)} \in [F(x) - 1/2^{l(x)}, F(x)), x \in A$$

and since $2^{-l(x)} \le p(x)/2$, it follows that

$$[\bar{F}(x)]_{l(x)} \in [F(x) - 1/2^{l(x)}, F(x)) \subset [F(x) - p(x)/2, F(x)) \subset (F(x-1), F(x)), x \in A$$

In particular, the numbers $[\bar{F}(x)]_{l(x)}, x \in A$, all fall in non-overlapping intervals and hence are all distinct. We wish to show that these codewords, in addition, satisfy the no prefix condition. Before proving that we observe that

$$\sum_{x \in A} l(x) p(x) = \sum_x p(x)([-log(p(x))] + 1) \le \sum_x p(x)(-log(p(x)) + 2) = H(p) + 2$$

Now observe that we also have

$$[\bar{F}(x)]_{l(x)} + 1/2^{l(x)} \le \bar{F}(x) + 1/2^{l(x)} \le \bar{F}(x) + p(x)/2$$

$$= F(x)$$

and hence

$$[\bar{F}(x)]_{l(x)} \subset [[\bar{F}(x)]_{l(x)}, [\bar{F}(x)]_{l(x)} + 1/2^{l(x)})$$

$$\subset (F(x-1), F(x)], x \in A$$

and hence the intervals

$$[[\bar{F}(x)]_{l(x)}, [\bar{F}(x)]_{l(x)} + 1/2^{l(x)}), x \in A$$

are all non-overlapping. In order to show that the codewords $[\bar{F}(x)]_{l(x)}, x \in A$ form a no-prefix code, it suffices therefore to show that if there are two positive integers $l \leq m$ such that for some binary fractions

$$z = 0.z_1z_2...z_l, u = 0.u_1u_2...u_m$$

we have that

$$[z, z + 1/2^l) and [u, u + 1/2^m)$$

are non-overlapping , then z cannot be a prefix of u. For suppose z is a prefix of u. Then we can write

$$u = 0.z_1..z_lu_{l+1}...u_m$$

and in particular,

$$z \leq u$$

Since the above intervals are non-overlapping, it then follows that

$$z + 1/2^l < u$$

However,

$$u = z + u_{l+1}/2^{l+1} + .. + u_m/2^m \leq z + 1/2^{l+1} + ... = z + 1/2^l$$

which is a contradiction. Hence the claim is proved.

Remark: The converse is also true, namely if z, u are as above with $l \leq m$ and z is not a prefix of u. Then $[z, z + 1/2^l)$ and $[u, u + 1/2^m)$ are non-overlapping. For suppose they are overlapping. Then since $l \leq m$, the second interval has length no larger than the first and hence three cases are possible:

[a] $z \leq u < u + 1/2^m \leq z + 1/2^l$

Thus,

$$0.z_1...z_l \leq 0.u_1...u_m \leq 0.z_1...z_l + 1/2^{l+1} + ... + 1/2^m$$

The first inequality implies that if i is the smallest integer such that $z_i = 1$ and j is the smallest integer such that $u_j = 1$, then $i \leq j$. The second inequality on the other hand implies that $j \geq i$. Hence, $j = i$. Similarly, if i' is the second smallest integer such that $z_{i'} = 1$ and j' is the second smallest integer such that $u_{j'} = 1$, then the first inequality would imply that $i' \geq j'$ while the second inequality would imply that $j' \geq i'$. Continuing in this way, we would get $j' = i'$ etc and hence finally, we would conclude that z must be a prefix of u, a contradiction.

[b] $u < z < u + 1/2^m$

This option may be ruled out since we can assume that $z \leq u$ because the length l of z is $\leq$ the length m of u. We have then that

$$0.u_1...u_m < 0.z_1...z_l < 0.u_1...u_m111...$$

Let i be the smallest integer for which $u_i = 1$ and j the smallest integer for which $z_j = 1$. Then the first inequality implies that $i \geq j$ and the second one implies that $j \geq i$. Likewise for the second smallest integers for which the corresponding entries of z and u are one and so on. Continuing in this way, we deduce that z must be a prefix of u.

Remark: Note that we can have a situation in which z is not a prefix of u but $z = u + 1/2^m$. Then

$$0.z_1...z_l = 0.u_1...u_m 11111...$$

Let i be the last index for which $u_i = 0 (i \leq m)$. Then

$$0.u_1...u_m 111... = 0.u_1...u_{i-1}1$$

Hence the condition implies

$$0.z_1...z_l = 0.u_1...u_{i-1}1$$

which implies that $l = i$, $z_l = 1$, $z_k = u_k, k = 1, 2, ..., i - 1$. Indeed in this case, the required semi-open intervals are still non-overlapping and z is not a prefix of u.

[d] $u < z + 1/2^l < u + 1/2^m$.

This means that

$$0.u_1...u_m < 0.z_1..z_l 111... < 0.u_1...u_m 111...$$

Let i be the smallest integer for which $u_i = 1$ and j the smallest integer for which $z_j = 1$. Then the first inequality implies that $i \geq j$ and the second implies that $j \geq i$ so $i = j$. Likewise for the second smallest integer and so on. Since $l \leq m$, it follows then that $0.z_1..z_l = 0.u_1..u_l$, ie, z is a prefix of u, a contradiction.

12.16   Some control problems involving the theory of large deviations

[1] Let $X(n), n \geq 1$ be a discrete time Markov process with transition probability density $\pi_\theta(x, y)$ dependent upon a vector parameter θ. Thus,

$$P(X(n + 1) \in B | X(n) = x) = \int_B \pi_\theta(x, y) dy$$

The rate function for the empirical density

$$l_N(x) = N^{-1} \sum_{n=1}^{N} \delta(x - X(n))$$

of the process is well known and is given by

$$I_N(q) = sup_{u>0} \int log(\frac{u(x)}{\pi_\theta u(x)}) q(x) dx$$

where

$$\pi_\theta u(x) = \int \pi_\theta(x,y)u(y)dy = \mathbb{E}_\theta(u(X(n+1))|X(n) = x)$$

where $\mathbb{E}_\theta$ denotes conditional expectation w.r.t the transition probability density π_θ. Assume that θ is known except for a small perturbation $\delta\theta$:

$$\theta = \theta_0 + \delta\theta$$

We can then expand

$$\pi_\theta u(x) \approx \pi_0 u(x) + \delta\theta(k)\pi_{1k}u(x) + (1/2)\delta\theta(k)\delta\theta(m)\pi_{2km}u(x)$$

where

$$\pi_0 u(x) = \pi_{\theta_0} u(x),$$

$$\pi_{1k} = \frac{\partial \pi_\theta}{\partial \theta(k)}|_{\theta=\theta_0},$$

$$\pi_{2km} = \frac{\partial \pi_\theta}{\partial \theta(k)\partial \theta(m)}|_{\theta=\theta_0}$$

Then,

$$log(\pi_\theta u(x)/u(x)) \approx log(\pi_0 u(x)/u(x)) + \delta\theta(k)(\pi_{1k}u(x)/\pi_0 u(x))$$

$$+\delta\theta(k)\delta\theta(m)[\pi_{2km}u(x)/2\pi_0 u(x) - \pi_{1k}u(x)\pi_{2m}u(x)/2(\pi_0 u(x))^2]$$

This expansion is valid upto quadratic orders in $\delta\theta$. Summation over the repeated indices k, m is implied. So with this approximation,

$$I(q) = -inf_{u>0}[\int q(x)log(\pi_0 u(x)/u(x))dx + \delta\theta(k)$$

$$\int q(x)\phi_{1k}(x)dx + (1/2)\delta\theta(k).\delta\theta(m)\int q(x)\phi_{2km}(x)dx]$$

where

$$\phi_{1k}(x) = \pi_{1k}u(x)/\pi_0 u(x), \phi_{2km}(x)$$

$$= \pi_{2km}u(x)/\pi_0 u(x) - \pi_{1k}u(x)\pi_{2m}u(x)/(\pi_0 u(x))^2 ---(a)$$

We wish to design the parameters $\delta\theta$ so that the probability that the empirical distribution $l_N = (l_N(x))$ will take a given value $p_0 = (p_0(x))$ is maximized. This probability is approximately given by $exp(-N.I(p_0))$ and hence we seek to determine $\delta\theta$ so that $I(p_0)$ is a minimum. With the above approximation, $\delta\theta$ is chosen as

$$argmax_{\delta\theta} inf_{u>0}[\int p_0(x)log(\pi_0 u(x)/u(x))dx + \delta\theta(k)$$

$$\int p_0(x)\phi_{1k}(x)dx + (1/2)\delta\theta(k).\delta\theta(m)\int p_0(x)\phi_{2km}(x)dx]$$

$$= inf_{u>0}argmax_{\delta\theta}[\int p_0(x)log(\pi_0 u(x)/u(x))dx + \delta\theta(k)\int p_0(x)\phi_{1k}(x)dx$$

$$+(1/2)\delta\theta(k).\delta\theta(m)\int p_0(x)\phi_{2km}(x)dx] ---(b)$$

Note that $\phi_{1k}(x)$ and $\phi_{1km}(x)$ are functionals of $u(.)$ as is evident from (a). Setting the gradient in the above expression w.r.t $\delta\theta$ to zero gives us for a given u,

$$\int p_0(x)\phi_{1k}(x)dx + \sum_m \delta\theta(m)\int p_0(x)\phi_{2km}(x)dx = 0$$

This is a linear set of equations for $\delta\theta$ and solving them gives us $\delta\theta$ as a functional of u. This value of $\delta\theta$ is then substituted back int (b) and minimized over u to get the optimal u which is substituted into the expression for $\delta\theta$ to obtain the value of the control parameter perturbation $\delta\theta$.

Examples of controllers based on the ldp.
In continuous time, we consider a one dimensional diffusion process

$$dX(t) = f(X(t))dt + g(X(t))dB(t)$$

The generator of this diffusion is

$$L = f(x)d/dx + (a(x)/2)d^2/dx^2, a(x) = g(x)^2$$

The rate function for the empirical density of $X(.)$ is given by

$$I(q) = -inf_{u>0} \int q(x)(Lu(x)/u(x))dx$$

Now

$$\int q(x)Lu(x)dx/u(x) = \int [q(x)(u'(x)/u(x) + a(x)u''(x)/2u(x))]dx$$

$$= \int [q(x)u'(x)/u(x) - u'(x)((q(x)a(x))'/2u(x) - q(x)a(x)u'(x)/2u(x)^2)]dx$$

$$= \int ((q - (qa)'/2)v + qav^2/2)dx$$

where

$$v(x) = u'(x)/u(x)$$

This is easily minimized w.r.t v to give

$$qav + (q - (qa)'/2) = 0$$

Chapter 13

Examination Problems in Classical Information Theory

[1] In a relay channel, let Tx and Rx denote respectively the main transmitter and receiver. Let RRx denote the intermediate relay receiver and RTx the intermediate relay transmitter. Tx transmits x after encoding a message while RTx transmits x_1 after appropriate encoding. RRx receives y_1 while Rx receives y. Note that these signals received by both RRx and Rx are channel noise corrupted versions of some combinations of x, x_1. In other words, RRx and Rx both receive messages from Tx and RTx via after channel noise corruption. The relay channel is thus completely specified by the transition probabilities $p(y, y_1|x, x_1)$. We assume a special case of this channel namely a degraded relay channel in which

$$p(y, y_1|x, x_1) = p(y_1|x, x_1)p(y|x_1, y_1)$$

In such a channel, RRx receives signals directly from both Tx and RTx after channel noise corruption while Rx receives messages directly from Tx and RTx after noise corruption. Thus Rx does not receive signals directly from Tx. It receives signals indirectly from Tx via RTx and RRx. Note that we can write for such a degraded channel,

$$y_1 = f_1(x, x_1, w_1), y = f(x_1, y_1, w_2) = f(x_1, f(x, x_1, w_1), w_2)$$

where w_1, w_2 are channel noise processes. From this expression, knowing the channels f_1, f and the statistics of the noise (w_1, w), it is easy to calculate $p(y, y_1|x, x_1)$. Now prove the following statements:

[a] The relay random encoder RTx generates 2^{nR_0} iid codewords $x_1^n(s), s \in E_0$ with pdf $p(x_1^n) = \Pi_i p(x_{1i})$. Here, $E_0 = \{1, 2, ..., 2^{nR_0}\}$.

[b] For each $s \in E_0$, conditioned on the codeword $x^n(s)$ generate in the relay, the main transmitter Tx generates 2^{nR} iid codewords $x^n(w|s), w \in E$ with condition pdf
$$p(x^n|x_1^n(s)) = \Pi_i p(x_i|x_{1i}(s))$$

227

where
$$E = \{1, 2, ..., 2^{nR}\}$$

[c] A random partition of E into $|E_0| = 2^{nR_0}$ cells $S(s), s \in E_0$ is generated by assigning to each $w \in E$ an $s \in E_0$ with uniform distribution. In other words, if $s(w)$ denotes the element of E_0 assigned to $s \in E$, then $s(w), w \in E_0$ are iid random variables with each $s(w)$ being uniformly distributed over E_0. Then for any given $s \in E_0$, the bin $S(s') \subset E$ is defined by

$$S(s') = \{w \in E | s(w) = s'\}, s' \in E_0$$

It follows that $S(s'), s' \in E_0$ is a random partition of E with

$$\mathbb{E}|S(s')| = 2^{n(R-R_0)}$$

$$Var(|S(s')|) = 2^{nR}(2^{-nR_0})(1 - 2^{-nR_0}) = 2^{n(R-R_0)}(1 - 2^{-nR_0}), s' \in E_0$$

[d] At time $i - 1$, Tx sends w_{i-1} and at time i w_i. We denote by s_i, the bin index of w_{i-1}. Thus, assuming that the relay (RRx, RTx) has successfully decoded w_{i-1}, it can generate the bin index s_i with good accuracy. Note that $w_{i-1} \in S(s_i)$.

[e] At time block i, the relay transmitter sends $x_1(s_i)$ (ie $x_1^n(s_i)$) and at the same time block i, the main transmitter Tx sends $x(w_i|s_i)$ (ie, $x^n(w_i|s_i)$). The corresponding signals received in the ith time block by RRx and Rx are respectively $y_1(i), y(i)$. Thus, $(x_1^n(s_i), x^n(w_i|s_i), y_1^n(i), y^n(i))$ is jointly typical since it has the given distribution

$$p(x_1^n)p(x^n|x_1^n)p(y^n, y_1^n|x^n, x_1^n) = \Pi_j p(x_{1j})p(x_j|x_{1j})p(y_{1j}|x_{1j}, x_j)p(y_j|x_{1j}, y_{1j})$$

[f] The relay receiver RRx estimates w_i as that $w \in E$ for which

$$(x(w|s_i), x_1(s_i), y_1(i))$$

is typical. Note that here we are assuming that the relay receiver has a good estimate of w_{i-1} and hence of s_i (the bin number in which w_{i-1} falls). Note that both the main receiver and the relay receivers are in knowledge of the codes $s \to x_1^n(s)$ and $(w, s) \to x^n(w|s)$. Using the fact that $(x(w_i|s_i), x_1(s_i), y_1(i))$ is typical (ie jointly typical) and hence that $(x_1(s_i), y(i))$ is typical (and hence $x_1(s_i)$ typical and $y_1(i)$ is conditionally typical given $x_1(s_i)$) and that for $w \neq w_i$, conditioned on s_i and $x_1(s_i)$, $x(w|s_i)$ is independent of $y_1(i)$, show that the probability of the relay receiver not estimating w_i correctly converges to zero provided that

$$R < I(x : y_1|x_1)$$

Remark: if $(U^n, V^n) = \{(U_i, V_i)\}_{i=1}^n$ is typical, then so is U^n and V^n is conditionally typical given U^n. This is because since U^n is typical,

$$p(V^n|U^n) = p(V^n, U^n)/p(U^n) \approx 2^{-nH(U,V)}/2^{-nH(U)} = 2^{-nH(V|U)}$$

[g] Rx estimates the bin number s_i (to which w_{i-1} belongs) from his received signal $y(i)$ in the ith time block as that $s \in E_0$ for which $(x_1(s), y(i))$ is typical. Show using the fact that $(x_1(s_i), y(i))$ is typical and that if $s \neq s_i$ $x_1(s)$ is independent of $y(i)$ (because $y(i) = f(x_1(s_i), x(w_i|s_i), n)$ and that $(x_1(s_i), x(w_i|s_i))$ is independent of $x_1(s)$ for $s \neq s_i$ because $x_1(s_i)$ is independent of $x(s)$ and conditioned on $x_1(s_i)$, $x(w_i|s_i)$ is independent of $x_1(s)$ when $s \neq s_i$) that the probability of RRx making an error in estimating s_i correctly converges to zero provided that

$$R_0 < I(x_1 : y)$$

[h] Assume that $s_k, k \leq i$ have all been decoded correctly at Rx as in [g]. We've seen that this is true if $R_0 < I(x_1 : y)$. Rx constructs the bin $S(s_i)$ and knows that $w_{i-1} \in S(s_i)$. Rx also knows the codeword $x_1(s_{i-1})$ since s_{i-1} has also been correctly decoded by him. He then estimates w_{i-1} as that $w \in S(s_i)$ for which $(x(w|s_{i-1}), x_1(s_{i-1}), y(i-1))$ is typical or equivalently since $x_1(s_{i-1})$ is typical (by construction of the random codeword $x_1(s_{i-1})$ as having the pdf $p(x_1^n) = \Pi p(x_{1i})$) as that $w \in S(s_i)$ for which $(x(w|s_{i-1}), y(i-1))$ is conditionally typical given $x_1(s_{i-1})$. A decoding error occurs regarding w_{i-1} therefore if either $(x(w_{i-1})|s_{i-1}), y(i-1))$ conditioned on $x_1(s_{i-1})$ is not typical (the probability of which converges to zero) or else if $(x(w|s_{i-1}), y(i-1))$ is conditionally typical given $x_1(s_{i-1})$ for some $w \neq w_{i-1}, w \in S(s_i)$ (the probability of which converges to zero provided that $R - R_0 < I(x : y|x_1)$. This is because the number of elements in $S(s_i)$ behaves as $2^{n(R-R_0)}$ asymptotically on the logarithmic scale, ie, $n^{-1} log(|S(s_i)|) \to R - R_0$ and further that conditioned on $x_1(s_{i-1})$, $x(w|s_{i-1})$ is independent of $y(i-1)$ when $w \neq w_{i-1}$ since $y(i-1) = f(x(w_{i-1}|s_{i-1}), x_1(s_{i-1}), n))$. Hence, conclude that the rate region is given by

$$R_0 < I(x_1 : y), R - R_0 < I(x : y|x_1)$$

[2] Let X^n be iid with distribution either P_1 or P_0. We wish to test whether the distribution is P_1 or P_0. Define

$$A(n, \epsilon) = \{X^n : |n^{-1}.log(P_1(X^n)/P_0(X^n)) + D(P_0|P_1)| < \epsilon\}$$

Show that

$$P_0(A(n, \epsilon)) \to 1, P_0(A(n, \epsilon)^c) \to 0$$

Define

$$Z(n) = \{X^n : n^{-1}.log(P_1(X^n)/P_0(X^n)) > -D(P_0|P_1) + \epsilon\}$$

Show that

$$Z(n) \subset A(n, \epsilon)^c$$

and hence

$$P_0(Z(n)) \to 0$$

Show that

$$P_1(X^n) \le P_0(X^n).exp(-n(D(P_0|P_1) - \epsilon)), X^n \in Z(n)^c$$

and hence that

$$P_1(Z(n)^c) \le exp(-n(D(P_0|P_1) - \epsilon))$$

and hence that

$$limsup n^{-1}.log(P_1(Z(n)^c) \le -D(P_0|P_1) + \epsilon$$

Conversely, suppose $Z(n)$ is any region of A^n where A is the alphabet in which X_i takes values. Suppose

$$liminf P_0(Z(n)) = \alpha < 1$$

Then show that

$$P_1(Z(n)^c) \ge P_1(Z(n)^c \cap A(n, \epsilon)) \ge P_0(Z(n)^c \cap A(n, \epsilon)).exp(-n(D(P_0|P_1) + \epsilon))$$

$$\ge (P_0(Z(n)^c) - P_0(A(n, \epsilon)^c)).exp(n(D(P_0|P_1) + \epsilon))$$

and hence conclude that

$$liminf n^{-1} log(P_1(Z(n)^c) \ge -D(P_0|P_1)$$

What are the physical implications of these results ?

[3] Let X be a random vector taking values in $\mathbb{R}^n$. Prove that conditioned on $\mathbb{E}(XX^T) = K_X$, $H(X)$ attains its maximum value when $X = N(0, K_X)$. Using this result, deduce that for any random vector $X \in \mathbb{R}^n$, the following inequality holds:

$$|K_X| \ge (2\pi e)^{-1}.exp(2H(X)/n)$$

[4] Let a memoryless Gaussian multiple access channel be described by the encoders

$$w_1 \to X_1^n(w_1), w_1 = 1, 2, ..., 2^{nR_1},$$

$$w_2 \to X_2^n(w_2), w_2 = 1, 2, ..., 2^{nR_2}$$

and the channel

$$Y = a_1 X_1 + a_2 X_2 + Z$$

where $Z = N(0, N)$. Under the power constraints $\| X_1^n(w_1) \| \le nP_1, \| X_2^n(w_2) \| \le nP_2$, calculate the maximal optimal regions for the rate pair (R_1, R_2). Also prove that the probability distribution of the source symbols X_1, X_2 corresponding to this maximal optimal regions is that for which X_1, X_2 are independent Gaussian distributions $N(0, P_1), N(0, P_2)$.

Generalize this result to the case of $m \geq 2$ message inputs with power constraints:

$$w_j \to X_j^n(w_j), w_j = 1, 2, ..., 2^{mR_j}, j = 1, 2, ..., m,$$

$$\| X_j^n(w_j) \|^2 \leq nP_j, j = 1, 2, ..., m$$

with the channel

$$Y = \sum_{j=1}^{m} a(j)X_j + Z, Z = N(0, N)$$

In these expressions,

$$\| X^n \|^2 = \sum_{j=1}^{n} X_j^2, X^n = ((X_j))_{j=1}^n$$

[5] [a] Prove the converse of the Slepian-Wolf theorem for distributed source coding with two sources X, Y so that $(X^n, Y^n) = \{(X_i, Y_i)\}_{i=1}^n$ is iid bivariate. The distributed encoders have the form

$$X^n \to f_n(X^n) = I_n \in \{1, 2, ..., 2^{nR_1}\},$$

$$Y^n \to g_n(Y^n) = J_n \in \{1, 2, ..., 2^{nR_2}\}$$

and the decoders have the form

$$\hat{X}^n = h_n(I_n, J_n), \hat{Y}^n = k_n(I_n, J_n)$$

If the decoding error probability converges to zero, then show using Fano's inequality that

$$n^{-1}H(X^n, Y^n|I_n, J_n) \to 0$$

and then deduce that

$$R_1 \geq H(X|Y), R_2 \geq H(Y|X)$$

[b] Let U, V be independent r.v's $U = N(0, \sigma_U^2), V = N(0, \sigma_V^2)$. Let $X = aU + bV, Y = cU + dV$ with $a, b, c, d, ad - bc$ all non zero. Calculate the Slepian-Wolf rate region for distributed encoding of (U, V) and then for (X, Y). Which rate region is larger and why ?

[6] [a] Prove that conditioning reduces entropy, ie, $H(X|Y) \leq H(X)$. What is the physical meaning of this result in terms of knowledge and uncertainty ?
[b] If $X \to Y \to Z \to W$ is a Markov chain, then show that

$$I(Y : Z) \geq I(X : Y)$$

and interpret this result.

[c] If $\{X_n\}_{n \in \mathbb{Z}}$ is a stationary stochastic process, then show that

$$H(X_0|X_{-1}, X_{-2}, ..., X_{-n}) =$$

$H(X_n|X_{n-1}, ..., X_0)$ and that this is a decreasing function of n. Deduced that its limit

$$\bar{H}(X) = lim_{n \to \infty} H(X_n|X_{n-1}, ...X_0) = H(X_0|X_{-1}, X_{-2}, ...)$$

exists that

$$\bar{H}(X) = lim_{n \to \infty} \frac{H(X_n, X_{n-1}, ..., X_1)}{n}$$

[d] Let $X_n \in \mathbb{R}^p, n \in \mathbb{Z}$ be a zero mean stationary Gaussian p-vector valued process with power spectral density matrix $S(\omega) \in \mathbb{R}^{p \times p}$. Let $R(n)$ denote the $np \times np$ correlation matrix $((\mathbb{E}(X_{i+k}X_i^T)))_{0 \le k \le n-1}$. Show that $R(n)$ is also the correlation matrix of the $np \times 1$ vector

$$[X_n^T, X_{n-1}^T, ..., X_1^T]^T$$

Let $R_X(m) = \mathbb{E}(X_{n+m}X_n^T) \in \mathbb{R}^{p \times p}$. Show that $R(n)$ can be expressed in $p \times p$ block structure form

$$R(n) = ((R_X(i - j)))_{1 \le i,j \le n}$$

Prove using the definition of entropy of a random vector in terms of its pdf that

$$H(X_n, ..., X_1) = -\mathbb{E}[ln f_X(X_n, ..., X_1)] = (1/2)ln(|R(n)|) + (np/2)ln(2\pi e)$$

$$H(X_n|X_{n-1}, ..., X_1) = -\mathbb{E}(ln(f_X(X_n|X_{n-1}, ..., X_1))$$

$$= (1/2)ln(|R(n)|/|R(n-1)|) + (p/2)ln(2\pi e)$$

Show that the eigenvalue equation

$$R(n)v = \lambda v$$

can be expressed in block form as

$$\sum_{m=0}^{n-1} R_X(k - m)v(m) = \lambda v(k), k = 0, 1, ..., n - 1$$

By taking the DFT on both sides and assuming $R_X(k)$ to be periodic with period n, show that this eigenvalue equation can equivalently be expressed as

$$S(2\pi k/n)V(k) = \lambda V(k), k = 0, 1, ..., n - 1$$

wher $\{V(k)\}$ is the n-point DFT of $\{v(k)\}$. Hence, deduce that the product of all the eigenvalues of $R(n)$ assuming $R_X(k)$ to be n-periodic is given by $\Pi_{k=0}^{n-1} det(S(2\pi k/n))$, ie,

$$det(R(n)) \approx \Pi_{k=0}^{n-1} det(S(2\pi k/n)), n \to \infty$$

Note that in the limit $n \to \infty$, there is no loss of generality in assuming n-periodicity for an aperiodic sequence. Thus, deduce that the entropy rate of a p-vector valued stationary Gaussian process is given by

$$\bar{H}(X) = lim_{n \to \infty} (2n)^{-1} \sum_{k=0}^{n-1} log(S(2\pi k/n)) + (p/2)ln(2\pi e)$$

$$= (4\pi)^{-1} \int_0^{2\pi} log(det(S(\omega))d\omega + (p/2).log(2\pi e)$$

[e] Show that the relative entropy map

$$(p, q) \to D(p|q) = \sum_x p(x).log(p(x)/q(x))$$

for two probability distributions p, q on a given finite set A is jointly convex, ie, if p_1, p_2, q_1, q_2 are four probability distributions on A and $0 \le t \le 1$, then

$$D(tp_1 + (1-t)p_2|tq_1 + (1-t)q_2) \le tD(p_1|q_1) + (1-t)D(p_2|q_2)$$

[7] Let $\{X_i\}_{i=1}^\infty$ be iid r.v's with continuous probability density $f(x)$. For $\epsilon > 0$, define

$$T(n, \epsilon) = \{X^n : |n^{-1}.log(f_n(X^n)) + H(X)| < \epsilon\}$$

where

$$X^n = (X_i)_{i=1}^n, f_n(X^n) = \Pi_{i=1}^n f(X_i)$$

and

$$H(X) = - \int f(x).log(f(x))dx$$

Now prove the following statements:

[a]

$$ex[(-n(H(X) + \epsilon)) \le f_n(X^n) \le exp * (-n(H(X) - \epsilon)), \forall X^n \in T(n, \epsilon)$$

[b]

$$P(X^n \in T(n, \epsilon)) \ge 1 - \sigma^2/n\epsilon^2, \sigma^2 = Var(log(f(X_1)))$$

[c]

$$(1 - \sigma^2/n\epsilon^2)exp(n(H(X) - \epsilon)) \le Vol(T(n, \epsilon)) \le exp(n(H(X) + \epsilon))$$

where for any Borel subset E of $\mathbb{R}^n$,

$$Vol(E) = \int_E d^n X$$

is its Lebesgue volume.

[d]
$$lim_{\epsilon \downarrow 0} lim_{n \to \infty} \frac{log(Vol(T(n, \epsilon)))}{n} = H(X)$$

[e] Explain the concept of Shannon's noiseless source coding theorem for sequences continuous iid random variables using [b] and [d].

[8] Let $\{X_n\}$ be a stationary stochastic process. Prove along the following steps that the entropy rate

$$\bar{H}(X) = H(X_0|X_{-1}, X_{-2}, ...) = limn^{-1}H(X_n, ..., X_1)$$

of this process given the autocorrelation lags $R(k)\mathbb{E}(X_{n+k}X_n), k = 0, 1, ..., p$ is a maximum when $\{X_n\}$ is a Gauss-Markov process of order p, ie, when X_n satisfies the difference equation

$$X_n = -\sum_{k=1}^{p} a(k)X_{n-k} + Z_n, n \geq p$$

with $[X_0, ..., X_{p-1}]^T$ having the $N(0, \mathbf{R}_p)$ distribution independent of $Z_n, n \geq p$ which is iid $N(0, \sigma^2)$, where $a(1), ..., a(p), \sigma^2$ are obtained by solving the $p + 1$ linear equations

$$R(m) + \sum_{k=1}^{p} a(k)R(m - k) = \sigma^2 \delta[m], m = 0, 1, ..., p$$

Here,
$$\mathbf{R}_p = ((R(i - j)))_{1 \leq i, j \leq p}$$

[a] Let Y_n be a zero mean stationary Gaussian process with same autocorrelations as those of X_n. Then, show that

$$H(X_n, ..., X_1) \leq H(Y_n, ..., Y_1) \forall n \geq 1$$

[b] Show that

$$H(Y_n, Y_{n-1}, ..., Y_1) \leq \sum_{k=1}^{n} H(Y_k|Y_{k-1}, ...Y_1)$$

and
$$\bar{H}(Y) = limn^{-1}H(Y_n, ..., Y_1) = H(Y_0|Y_{-1}, Y_{-2}, ...)$$
$$\leq H(Y_0|Y_{-1}, ..., Y_{-p})$$

with equality iff Y_n is Gauss-Markov of order p. Show further that $H(Y_0|Y_{-1}, ..., Y_{-p})$ is completely determined by $R(k), k = 0, 1, ..., p$.

[9] State and prove Fano's inequality and give its application in proving the converse part of Shannon's noisy coding theorem.

[10] let A be an alphabet and let $(X_i, Y_i, Z_i), i = 1, 2, ...$, be iid random vectors with values in $A \times A \times A$. For $a, b, c \in A$, let $N(a, b, c | X^n, Y^n, Z^n)$ denote the number of times that a, b, c appears at the corresponding positions of X^n, Y^n, Z^n, ie,

$$N(a, b, c) = N(E)$$

where

$$E = \{i : X_i = a, Y_i = b, Z_i = c, 1 \leq i \leq n\}$$

Let $\epsilon > 0$ and let

$$A_{XYZ}(n, \epsilon) = \{(X^n, Y^n, Z^n) :$$

$$|N(a, b, c | X^n, Y^n, Z^n)/n - p_{XYZ}(a, b, c)| < \epsilon \forall a, b, c \in A\}$$

Also put

$$A_{YZ}(n, \epsilon) = \{(Y^n, Z^n) : |N(b, c | Y^n, Z^n)/n - p_{XYZ}(a, b, c)| < \epsilon \forall b, c \in A\}$$

If $X_i \to Y_i \to Z_i$ is a Markov chain ie, if

$$p_{XYZ}(a, b, c) = p_X(a) p_{Y|X}(b|a) p_{Z|Y}(c|b)$$

then show that $Z_i \to Y_i \to X_i$ is also a Markov chain and hence deduce that $(Y^n, Z^n) \in A_{YZ}(n, \epsilon)$ implies

$$(X^n, Y^n, Z^n) \in A_{XYZ}(n, \delta)$$

where $\delta = \delta(\epsilon) \to 0$ as $\epsilon \to 0$. In other words show that

$$N(a, b, c | X^n, Y^n, Z^n)/n \to p_{XYZ}(a, b, c)$$

For proving this result, you must apply the law of large numbers to the idenitity

$$N(a, b, c | X^n, Y^n, Z^n)/n = (N(a | X^n_{b,c})/N(b | Y^n_c))(N(b | Y^n_c)/N(c | Z^n))(N(c | Z^n)/n)$$

and use the Markovianity $Z \to Y \to X$ to deduce that

$$N(c | Z^n)/n \to p_Z(c),$$

$$N(b | Y^n_c)/N(c | Z^n) \to p_{Y|Z}(b|c),$$

$$N(a | X^n_{b,c})/N(b | Y^n_c) \to p_{X|Y}(a|b)$$

The notation is that Y^n_c is that segment of Y^n in for which c occurs in the corresponding positions of Z^n and $X^n_{b,c}$ is that segment of X^n in which (b, c) occurs in the corresponding positions of (Y^n, Z^n):

$$Y^n_c = \{Y_i : Z_i = c\}, X^n_{b,c} = \{X_i : Y_i = b, Z_i = c\}$$

13.1 Converse part of the achievability result for a multiterminal network

Consider a newtork with m nodes numbered $E = \{1, 2, ..., m\}$ where each node can transmit as well as receive messages and decode them. The ith node wishes to send a message

$$W^{(ij)} \in E_{ij} = \{1, 2, ..., 2^{nR_{ij}}\}, i, j = 1, 2, ..., m$$

to the jth node. The set of all messages sent by the ith node is therefore

$$W^i = \{W^{ij} : j = 1, 2, ..., m\}$$

The encoded message sent by node i at time k is of the general form

$$X^i(k) = X^i_k(W^i, Y^i(l), l = 1, 2, ..., k-1) = X^i_k(W^{ij},$$
$$_i \qquad\qquad j = 1, 2, ..., m, Y^i(l), j \le k-1)---(1)$$

In this expression, Y (l) is the message received by node i due to transmission by all the nodes $j \ne i$ after taking into account channel noise which is assumed to be iid in time. The form (1) of the encoded signal at time k finally transmitted by node i is thus a function of all the messages $W^i = (W^{ij} : j = 1, 2, ..., m)$ that node i wishes to transmit and also of all the output signals that node i receives before time k. In other words, we allow for a causal output feedback mechanism at each node in generating the signal to be finally transmitted at any given time. Let S be a set of nodes and $S^c = \{1, 2, ..., m\} - S = E - S$ its complement. Let $\{R^{ij} : i, j \in E\}$ be a set of achievable rates, ie, rates for which there exist feedback codes of the above form which assure asymptotically zero error of decoding probability as $n \to \infty$. Note that the decoders at each node are assumed to have the following form:

$$\hat{W}^{ij} = \hat{W}^{ij}(Y^j, W^j) = W^{ij}(Y^j, W^{jl}, l = 1, 2, ..., m)$$

where by X^i, we mean all its time samples $\{X^i(k) : k \le n\}$ and by Y^i, likewise, we mean $\{Y^i(k) : k \le n\}$ and $\{\hat{X}^i(k) : k \le n\}$. Since the decoding error probability converges to zero as $n \to \infty$, it follows from Fano's inequality that

$$H(W^{ij} | Y^j, W^j) \to 0$$

Note that the decoder at node j, while decoding the message W^{ij} transmitted by node i to him, is allowed to make use of only the signal Y^j received by him as well as all the messages W^j transmitted by him to the other nodes. From the above convergence, it is easily deduced that

$$H(W^T | Y^{S^c}, W^{T^c})/n \to 0$$

where
$S \subset E$, and $T = S \times S^c$, so that

$$W^T = W^{S,S^c} = \{W^{ij} : i \in S, j \in S^c\}$$

is the set of all messages transmitted by the nodes in S to nodes in S^c,

$$W^{T^c} = W^{S^c,S} \cup W^{S^c,S^c} = W^{S^c} = W^{S^c,E}$$

is the set of all messages transmitted from S^c to some node of E. Note that the decoders for all messages sent from S to S^c are located in S^c and these decoders only use the received noisy signals $Y^{S^c} = (Y^j, j \in S^c)$ and the messages sent from S^c to some node in E. We can also therefore safely write the condition that the decoding error probability goes to zero as

$$H(W^T|Y^{S^c}, W^{T^c}) = H(W^{S,S^c}|Y^{S^c}, W^{S^c})/n \to 0$$

Note that Y^E is a function of X^E and iid channel noise with the channel consisting of all the $m(m-1)$ directed paths joining the different nodes. Now we assume that the $W^{ij}, i, j \in E, i \neq j$ are all independent random variables with W^{ij} being uniformly distributed over E_{ij}. Then for any subset $S \subset E$, we have with $T = S \times S^c$,

$$nR^{S,S^c} = n \sum_{i \in S, j \in S^c} R^{ij} \leq H(W^T) = H(W^T|W^{T^c})$$

since by hypothesis the messages transmitted by different nodes are independent and hence W^T is independent of W^{T^c}. Note that

$$W^T = W^{S,S^c}, W^{T^c} = W^{S^c} = W^{S,S} \cup W^{S,S^c} = W^{S,E}$$

Now,

$$H(W^T|W^{T^c}) = H(W^T|Y^{S^c}, W^{T^c}) + I(W^T : Y^{S^c}|W^{T^c})$$

and we have already seen that since the decoding error probability converges to zero,

$$H(W^T|Y^{S^c}, W^{T^c})/n \to 0$$

Further,

$$I(W^T : Y^{S^c}|W^{T^c}) = H(Y^{S^c}|W^{T^c}) - H(Y^{S^c}|W^T, W^{T^c})$$

$$= \sum_k (H(Y^{S^c}(k)|Y^{S^c}(j), j \leq k-1, W^{T^c}) - H(Y^{S^c}(k)|Y^{S^c}(j), j \leq k-1, W^T, W^{T^c}))$$

$$\sum_k (H(Y^S(k)|Y^S(j), j \leq k-1, W^S, X^S(k)) - H(Y^S(k)|Y^S(j), j \leq k-1, W^T, W^T))$$

$$=$$

(because $X^{S^c}(k)$ is a deterministic function of $(Y^{S^c}(j), j \leq k-1, W^{S^c})$

$$\leq \sum_k (H(Y^{S^c}(k)|X^{S^c}(k)) - H(Y^{S^c}(k)|Y^{S^c}(j), j \leq k-1, W^T, W^{T^c}))$$

(because conditioning reduces entropy)

$$= \leq \sum_k (H(Y^{S^c}(k)|X^{S^c}(k)) - H(Y^{S^c}(k)|Y^{S^c}(j), j \leq k-1, W^T, W^{T^c}, X^S(k), X^{S^c}(k)))$$

(because conditioning reduces entropy)

$$= \sum_k (H(Y^{S^c}(k)|X^{S^c}(k)) - H(Y^{S^c}(k)|X^S(k), X^{S^c}(k)))$$

(because $Y^{S^c}(k)$ is a function of $X^E(k) = (X^S(k), X^{S^c}(k))$ and channel noise at time k and therefore since the channel noise is iid, and independent of $X^E(k)$, it follows that conditioned on $X^E(k)$, $Y^{S^c}(k)$ is independent of $Y^{S^c}(j), j \leq k - 1, W^T, W^{T^c}$). Now define a r.v Q that assumes values $1, 2, ..., n$ with probability $1/n$ each and is independent of $X^E(k), Y^E(k), k = 1, 2, ..., n$, ie, is independent of $X^E(k), N^E(k), k = 1, 2, ..., n$ where N^E is the network channel noise. Then, we get from the above that

$$I(W^T : Y^{S^c}|W^{T^c}) \leq H(Y^{S^c}(Q)|X^{S^c}(Q), Q) - H(Y^{S^c}(Q)|X^S(Q), X^{S^c}(Q), Q)$$

$$\leq H(Y^{S^c}(Q)|X^{S^c}(Q)) - H(Y^{S^c}(Q)|X^S(Q), X^{S^c}(Q))$$

because conditioning reduces entropy and secondly, conditioned on $X^E(Q)$, $Y^{S^c}(Q)$ depends only on the iid network channel noise and is hence independent of Q. Specifically, given the event

$$\{Q = i, X^E(Q) = x\} = \{Q = i, X^E(i) = x\}$$

the distribution of $Y^E(Q)$ equals that of $Y^E(i)$ whose distribution under this conditioning, depends only on x and the distribution of the channel noise $N^E(i)$ which owing to the iid nature of the channel noise, is independent of i.

13.2   More Examination Problems in information theory

[1] Consider a degraded broadcast channel with input x and successive outputs y_1, y_2, so that $x \to y_1 \to y_2$ forms a Markov chain. The source and channel are thus characterized by the joint pdf

$$p(x, y_1, y_2) = p(x)p(y_1|x)p(y_2|y_1)$$

We introduce an auxiliary r.v U so that the joint distribution of (U, X, Y_1, Y_2) becomes

$$p(u, x, y_1, y_2) = p(u)p(x|u)p(y_1|x)p(y_2|y_1)$$

ie,

$$u \to x \to y_1 \to y_2$$

forms a Markov chain. Choose iid words $U^n(w_2), w_2 = 1, 2, ..., 2^{nR_2}$ with pdf $p(U^n) = \Pi p(u_i)$. For each w_2 given $U^n(w_2)$, choose iid words $X^n(w_1, w_2), w_1 = 1, 2, ..., 2^{nR_1}$ with pdf $p(X^n|U^n(w_2)) = \Pi_i p(x_i|u_i(w_2))$. Thus the rate of transmission, or code rate is R_1 for this random code. We can look upon this degraded broadcast channel as transmitting two messages w_1, w_2 at rates R_1 and R_2 respectively. If (w_1, w_2) is the message pair to be transmitted, then, the transmitted codeword over the broadcast channel is $X(w_1, w_2)$ and the received signals are respectively $(Y_1(w_1, w_2), Y_2(w_1, w_2))$. Assume that channel noise is iid and independent of all the sources. Then show that $(X(w_1, w_2), Y_1(w_1, w_2), Y_2(w_1, w_2))$, $w_1 = 1, 2, ..., 2^{nR_1}, w_2 = 1, 2, ..., 2^{nR_2}$ are $2^{n(R_1+R_2)}$ independent random vectors. As- sume that $(1, 1)$ is the transmitted message pair.

Now decode w_2 at receiver two as that w_2 for which $(U(w_2), Y_2(1, 1))$ is typical. Explain the fact that a decoding error occurs here if either $(U(1), Y_2(1, 1))$ is not typical or else if $(U(w_2), Y_2(1, 1))$ is typical for some $w_2 \neq 1$. Show that the decoding error probability at receiver two converges to zero if $R_2 < I(U : Y_2)$.

Now decode w_1 at receiver one as that w_1 for which $(U(w_2), X(w_1, w_2), Y_1(1, 1))$ is typical for some w_1. Explain the fact that a decoding error occurs here implies one of the following: (a) $(U(w_2), Y_1(1, 1))$ is typical for some $w_2 \neq 1$. Show that the probability of this happening converges to zero if $R_2 < I(U : Y_1)$. Show that degradedness of the broadcast channel, ie, $X \to Y_1 \to Y_2$ is a Markov chain implies $I(U : Y_1) \geq I(U : Y_2)$ and hence the condition $R_2 < I(U : Y_2)$ implies that this error probability is guaranteed to converge to zero. (b) $(U(1), X(w_1, 1), Y(1, 1))$ is typical for some $w_1 \neq 1$. Show that the probability of this happening converges to zero if $R_1 < I(X : Y_1|U)$. Combine these results to state the achievability region for a degraded broadcast channel.

[2] We wish to compress an iid source sequence X^n using available side information Y^n. Assume $(X^n, Y^n) = ((X_i, Y_i))_{i=1}^n$ is an iid bivariate sequence. Encode X^n at rate R_2 to $f_n(X^n) = i = 1, 2, ..., 2^{nR_1}$. The decoder has access to the code $f_n(X^n)$ and the side information Y^n in the form of another codeword $g_n(Y^n) = i = 1, 2, ..., 2^{nR_2}$ and thus decodes X^n as $\hat{X}^n = h_n(i, s) = h_n(f_n(X^n), g_n(Y^n))$. We wish to select the endoders f_n, g_n and the decoder h_n so that the decoding error probability $P(\hat{X}^n \neq X^n)$ converges to zero simultaneously keeping R_1, R_2 minimal. Any such R_1, R_2 which leads to zero asymptotic decoding error probability is called achievable. Show that if there exists a r.v. U such that $X \to Y \to U$ forms a Markov chain then the achievable region is $R_1 > H(X|U), R_2 > I(U : Y)$. Your proof can be based along the following lines:

[a] Generate iid random vectors $U^n(s), s = 1, 2, ..., 2^{nR_2}$ with pdf $p(U^n) = \Pi p(u_i)$. Partition the X^n sequences into 2^{nR_1} bins uniformly in an iid way, ie define iid uniform random variables $i(X^n) \in \{1, 2, ..., 2^{nR_1}\}, X^n \in A^n$. Let $B(j)$ denote the set of X^n for which $i(X^n) = j$. Show that if E is any set, then the average number of elements in $E \cap B(j)$ for any fixed j equals $|E|/2^{nR_1}$.

[b] If $(Y^n, U^n(s))$ is typical for some $s = 1, 2, ..., 2^{nR_2}$, then encode Y^n as the minimum of all such s. If $(Y^n, U^n(s))$ is not typical for any s, then encode Y^n as some fixed s_0. Thus unique decoding of Y^n from s fails in the latter case. Show that the probability of this happening converges to zero if $R_2 > I(U : Y)$.

[c] Assuming that $R_2 > I(U : Y)$, we can thus assume that $(Y^n, U^n(s))$ is typical and hence since $X \to Y \to U$ is a Markov chain, $(X^n, U^n(s))$ is typical. Encode X^n as i where $i = i(X^n)$, ie, i is the unique index for which $X^n \in B(i)$.

[d] The decoder has the pair of indices (i, s) available to him. He decodes X^n as that sequence $X'^n \in B(i)$ for which $(X'^n, U^n(s))$ is typical. Since we have already seen that $(X^n, U^n(s))$ is typical for the true sequence X^n, a decoding error can occur only when there is some $X'^n \neq X^n, X'^n \in B(i)$ for which $(X'^n, U^n(s))$ is typical. Show that the probability of this happening is bounded above by $|A_X \cap B(i)| . 2^{-nI(U:X)}$ which for large n equals $(|A_X|/2^{nR_1}) 2^{-nI(U:X)} = 2^{-n(R_1 + H(X) - I(U:X))} = 2^{-n(R_1 - H(X|U))}$ and it converges to zero if $R_1 > H(X|U)$. Here, A_X denotes the set of sequences X'^n which are typical (corresponding to the probability law $p(X^n) = \Pi p(x_i)$).

[3] Consider the problem of compressing X^n data with allowed distortion D and with the decoder allowed to make use of side information. The iid data with side information is $(X^n, Y^n) = \{(X_i, Y_i)\}_{i=1}^n$ where X_i is correlated with Y_i. Y^n forms the available side information. Let d be a metric on the X-space. Let

$$R_1 = I(X : W) + \epsilon_1, R_2 = I(X : W) - I(Y : W) + \epsilon_2$$

Let

$$D = \mathbb{E}(d(X, f(Y, W))$$

where (X, Y, W) has the distribution of any one sample of the iid vectors $(X_i, Y_i, W_i), i = 1, 2, ..., n$. Choose iid data $W^n(s), s = 1, 2, ..., 2^{nR_1}$ with pdf $p(W^n) = \Pi p(w_i)$. Encode X^n as the minimal s for which $(X^n, W^n(s))$ is typical. Show that the probability of there being no such s converges to zero if $R_1 > I(X : W)$ which we have assumed to satisfy. Show that $(Y^n, W^n(s))$ is typical as a consequence of the assumption that $Y \to X \to W$ is a Markov chain. Now assign to each $s' = 1, 2, ..., 2^{nR_1}$ a number $i = 1, 2, ..., 2^{nR_2}$ in an iid random uniform way. Let $B(j)$ denote the set of all s' which have been assigned the number j. Show that $B(j), j = 1, 2, ..., 2^{nR_2}$ is a random partition of $\{1, 2, ..., 2^{nR_1}\}$ with $\mathbb{E}(|B(j)|) = 2^{n(R_1 - R_2)}$ for any given j and show further that $Var(|B(j)|) = 2^{n(R_1 - R_2)}|(1 - 2^{-nR_2})$ and that this fact implies that $n^{-1} log(|B(j)|) - (R_1 - R_2)$ converges to zero almost surely as $n \to \infty$. Now choose that $i = 1, 2, ..., 2^{nR_2}$ for which $s \in B(i)$. In this way X^n has been assigned a unique $i \in \{1, 2, ..., 2^{nR_2}\}$ with probability converging to one as $n \to \infty$. The decoding process involves decoding from i, the value s and hence recovering X^n. This decoding process involves selecting that $s' \in B(i)$ for which $(Y^n, W^n(s'))$ is typical. s' is then the desired estimate of s. Explain the fact that a decoding error will occur only when there exists an $s' \neq s$ in $B(i)$ such that $(Y^n, W^n(s'))$ is typical. Show that the probability of such an event taking place is bounded above by $2^{n(R_1 - R_2)} . 2^{-nI(Y:W)}$ and that this probability will converge to zero when $R_2 > R_1 - I(Y : W) > I(X : W) - I(Y : W)$. Show further that the typicality of $(X^n, W^n(s))$ and the Markovianity of $Y \to X \to W$ implies the typicality of $(Y^n, X^n, W^n(s))$ and hence this triplet has the required distribution, namely $\Pi_i p(y_i, x_i, w_i)$ and hence by the weak law of large numbers, its average distortion $n^{-1} \sum_{i=1}^n d(x_i, f(y_i, w_i)) \approx D$. Conclude the result that by encoding at the rate of $I(X : W) - I(Y : W)$ the given data X^n as $n \to \infty$, we are able to achieve data compression with retrieval within the limits of the given distortion D.

More examination problems on information theory

[1] [a] State and prove Fano's inequality.

[b] Give an application of Fano's inequality to proving the converse part Shannon noisy coding theorem for discrete memoryless channels, namely if the error probability for a sequence of codes of increasing length converges to zero, then the rate of information transmission must be smaller than the channel capacity.

[2] Let A, B be finite alphabets. Let $\{(X_i, Y_i)\}_{i=1}^n$ be an iid bivariate sequence. For $\epsilon > 0$, define the Bernoulli typical sets

$$A_X(n, \epsilon) = \{X^n \in A^n : |N(a|X^n)/n - p_X(a)| < \epsilon \forall a \in A\}$$

and likewise $A_Y(n, \epsilon)$,

$$A_{XY}(n, \epsilon) = \{(X^n, Y^n) : |N(a, b|X^n, Y^n)|/n - p_{XY}(a, b)| < \epsilon \forall a \in A, b \in b\}$$

and conditionally Bernoulli typical set

$$A_{Y|X}(n, \epsilon) = \{Y^n \in B^n : (X^n, Y^n) \in A_{XY}(n, \epsilon)\}$$

Prove that

[a] $p_X(A_X(n, \epsilon)) \to 1$, [b] $p_{XY}(A_{XY}(n, \epsilon)) \to 1$,

[c] For large n, $|A_X(n, \epsilon)| \in [2^{-n(H(X)+\delta)}, 2^{-n(H(X)+\delta)}]$ where $\delta = \delta(\epsilon) \to 0$ as $\epsilon \to 0$, [d] $p_{Y|X}(Y^n|X^n) \approx 2^{-nH(Y|X)}$ for $Y^n \in A_{Y|X}(n, \epsilon), X^n \in A_X(n, \epsilon)$, [e] If $\tilde{Y}^n$ has the same iid distribution as Y^n but is independent of X^n, then $Pr(\tilde{Y}^n \in A_{Y|X}(n, \epsilon)|X^n) \approx 2^{-nI(Y:X)}$.

[3] [a] A Gaussian discrete memoryless multiple access channel has two inputs X_1, X_2 and output Y. The channel characteristics per symbol is defined by $Y = a_1 X_1 + a_2 X_2 + Z$ where the channel noise Z is $N(0, N)$. Calculate the rate region for (R_1, R_2).

[b] If X, Y are independent and we define $U = a_1 X + a_2 Y, V = b_1 X + b_2 Y$, will the Slepian-Wolf capacity region for distributed encoding of the iid bivariate source (U^n, V^n) be greater than or lesser than that of (X^n, Y^n). Explain your result.

[4] [a1] Prove the inequality

$$\sum_j a(j) log(\frac{a(j)}{b(j)}) \geq (\sum_j a(j)).log(\frac{\sum_j a(j)}{\sum_j b(j)})$$

for positive numbers $\{a(j), b(j)\}_{j=1}^n$. Using this, deduce joint convexity of the relative entropy

$$D(p|q) = \sum_i p(i) log(p(i)/q(i))$$

between two probability distributions p, q on the same set.

[a2] State and prove the data processing inequality for a four stage Markov chain and give one application of this.

[b] Explain how in a relay system, by transmitting bin indices at lower rate from the relay terminal, how one is able to achieve a higher capacity. State the difference between a relay channel and a physically degraded relay channel.

[5] [a] Given five symbols in a source with probability distribution

$$p(1) > p(2) > p(3) > p(4) > p(5)$$

such that

$$p(3) > p(4)+p(5), p(1) > p(4)+p(4)+p(5) > p(2), p(2)+p(3)+p(4)+p(5) > p(1)$$

write down the optimal Huffman code.

[b] Let $\{\mathbf{X}_i\}$ be iid $N(0, \mathbf{K}_X)$ random vectors in $\mathbb{R}^d$ with the eigenvalues of $\mathbf{K}_X$ being $P_1, ..., P_d$. Show that the rate distortion function $R(D)$ for $\mathbf{X}$ is given by

$$R(D) = (1/2)min_{\{D_i\}} \sum_{i=1}^{D} max(log(P_i/D_i), 0)$$

where the minimum is over all positive $\{D_i\}$ with $\sum_{i=1}^{d} D_i = D$. Using Lagrange multipliers, calculate the optimal values of $\{D_i\}$. Show that these are given by $D_i = \lambda.\theta(P_i - \lambda)$ where $\theta(x)$ is the unit step function and λ is chosen so that $\sum_{i=1}^{d} \lambda.\theta(P_i - \lambda) = D$. Note that the distortion metric being used here is $d(\mathbf{X}, \hat{\mathbf{X}}) = \mathbb{E} \parallel \mathbf{X} - \hat{\mathbf{X}} \parallel^2$ where $\parallel . \parallel$ is the Euclidean norm.

Also explain how rate distortion theory gives more compression than compression without distortion and how rate distortion with side information gives even more compression. Do this by comparing the compression values in all the three cases in terms of entropy and mutual information.

Problems in Statistical Signal Processing

[1] A uniform linear array consists of 3 sensors receiving signals from two sources. Assume the source directions to be $\theta_k, k = 1, 2$. Show that if the three sensors are located at points $0, d, 2d$ along the z-axis then the array steering vector along the direction θ is given by

$$\mathbf{e}(\theta) = [1, exp(jKd.cos(\theta)), exp(jK2d.cos(\theta))]^T$$

where $K = \omega/c$ with ω the centre frequency of the source signals and c the velocity of the electromagnetic signals. Show that the array signal vector can be expressed as

$$\mathbf{x}(t) = \mathbf{E}(\theta_1, \theta_2)\mathbf{s}(t) + \mathbf{w}(t)$$

where
$$\mathbf{s}(t) = [s_1(t), s_2(t)]^T$$
is the source signal lowpass envelope vector and $\mathbf{w}(t)$ is the sensor noise lowpass envelope vector. Assuming the noise to be white in both space and time and $s(t), \mathbf{w}(t)$ to be uncorrelated Gaussian signals with

$$Cov(\mathbf{w}(t)) = \sigma_w^2 \mathbf{I}_3, Cov(\mathbf{s}(t)) = \begin{pmatrix} P_1 & P_{12} \\ P_{12} & P_2 \end{pmatrix}$$

Derive

[a] The maximum likelihood estimator of $\theta_k, k = 1, 2$ from the iid measurements $\mathbf{x}(t_k), k = 1, 2, ..., M$

[b] The MUSIC pseudo-spectral estimator of $\theta_k, k = 1, 2$ and the ESPRIT spectral estimator of θ_1, θ_2 assuming a shifted array with sensors at $d, 2d, 3d$ with same noise statistics. Take as the steering vector of the shifted array,

$$\mathbf{f}(\theta) = exp(jKd.cos(\theta))\mathbf{e}(\theta)$$

Assume ergodicity of the sources and noise so that time averaged covariances can be replaced by the true ensemble averaged covariances.

[2] Consider the nonlinear sequential 1-D data model

$$x[n] = a(n)\theta + \epsilon.b(n)\theta^2 + v[n], n = 1, 2, ...$$

where $\{v[n]\}$ is iid $N(0, \sigma_v^2)$

[a] Calculate the maximum likelihood estimator $\hat{\theta}[N]$ of θ based on the measurements $x[n], 1 \le n \le N$ upto $O(\epsilon^2)$ using perturbation theory. How will you cast this estimator in time recursive form ?

[b] Calculate the mean and variance of $\hat{\theta}[N]$ upto $O(\epsilon^2)$ and also the Cramer-Rao lower bound on the variance of an unbiased estimator of θ.

[3] A signal $x[n] = s[n] + w[n]$ is such that s and w are uncorrelated processes, $s[n]$ has long range correlations $R_{ss}[\tau]$ while $w[n]$ has short range correlations $R_{ww}[\tau]$. In the line enhancement technique, we predict $x[n]$ linearly based on $x[n - D - k], k = 1, 2, ..., p$ where the delay D is large enough to guarantee near uncorrelatedness of $w[n]$ and $w[n - D - k]$ but small enough to guarantee that $s[n]$ is strongly correlated with $s[n - D - k]$. Explain how the predicted MMSE data for $x[n]$ will be a good estimate $\hat{s}[n]$ of $s[n]$. Derive a formula for the prediction error energy $\mathbb{E}(s(n) - \hat{s}(n))^2$ assuming $R_{ss}[\tau] = \sigma_s^2.exp(-a|\tau|), R_{ww}[\tau] = \sigma_w^2 exp(-b|\tau|)$ where $a << b$.

[4] Consider the time series model

$$x[n] = ax[n - 1] + w[n]$$

where $w[n]$ is a zero mean Gaussian process with autocorrelation $R_{ww}[\tau] = \sigma_w^2 exp(-a|\tau|)$.

[a]Show that $w[n]$ can be made to satisfy the difference equation

$$w[n] = cw[n-1] + v[n]$$

where $v[n]$ is white Gaussian noise and c is appropriately chosen. Write down the optimal causal Wiener filter for estimating $x[n]$ based on $y[n-k], k \geq 0$ where $y[n] = x[n] + \epsilon[n]$ with $\epsilon[n]$ white Gaussian with variance σ_ϵ^2.

[b] Write down the Kalman filer algorithm for estimating $(x[n], w[n])^T$ based on $y[n-k], k \leq 0$. Compare the this estimate in the steady state with the optimal causal Wiener filter.

[5] Write short notes on any two of the following.

[a] Convergence analysis of the LMS algorithm.

[b] Adaptive echo cancellation in telephone lines.

[c] Adaptive noise cancellation.

[d] Matrix inversion lemma with applications to extended Kalman filtering.

[e] Time and order recursive RLS lattice algorithm.

For Product Safety Concerns and Information please contact our EU
representative GPSR@taylorandfrancis.com
Taylor & Francis Verlag GmbH, Kaufingerstraße 24, 80331 München, Germany

www.ingramcontent.com/pod-product-compliance
Lightning Source LLC
Chambersburg PA
CBHW060247220326
41598CB00027B/4018